Advanced Classical and Quantum Probability Theory with Quantum Field Theory Applications

Advanced Classical and Quantum Probability Theory with Quantum Field Theory Applications

Harish Parthasarathy
Professor
Electronics & Communication Engineering
Netaji Subhas Institute of Technology (NSIT)
New Delhi, Delhi-110078

CRC Press
Taylor & Francis Group
Boca Raton London New York

CRC Press is an imprint of the
Taylor & Francis Group, an **informa** business

Manakin
PRESS

First edition published 2023
by CRC Press
2 Park Square, Milton Park, Abingdon, Oxon, OX14 4RN

and by CRC Press
6000 Broken Sound Parkway NW, Suite 300, Boca Raton, FL 33487-2742

CRC Press is an imprint of Informa UK Limited

British Library Cataloguing-in-Publication Data
A catalogue record for this book is available from the British Library

ISBN: 9781032405124 (hbk)
ISBN: 9781032405148 (pbk)
ISBN: 9781003353430 (ebk)

DOI: 10.4324/9781003353430

Typeset in Arial, BookManOldStyle, CourierNewPS, MS-Mincho, Symbol, and TimesNewRoman
by Manakin Press, Delhi

Manakin
PRESS

Preface

The contents of this book are based on three undergraduate and postgraduate courses taught by the author on Matrix theory, probability theory and antenna theory over the past several years. The portion on matrix theory covers basic linear algebra including quotient vector spaces, variational principles for computing eigenvalues of a matrix, primary and jordan decomposition theorems for non-diagonable matrices, simultaneous triangulability, and basic matrix decomposition theorems useful in statistics, signal processing and control. It also covers Lie algebra theory culminating the celebrated root space decomposition introduced by E.Cartan of a semisimple Lie algebra in terms of Cartan subalgebras and root vectors and also some interesting topics in control theory like controllability of partial differential equations of mathematical physics including Maxwell's equations, Dirac equation and their quantum versions. By control of a quantum system comprising electrons, positrons and photons described by the second quantized Maxwell and Dirac equations, we mean the design of classical control fields like current and electromagnetic fields so that the resulting radiation pattern genererated by the electron-positron field will have space-time moments that provide a good match to a given set of moments. Some discussion on large deviation theory in control has also been included involving designing control fields for pde's driven by weak stochastic noise so that the probability of deviation of the controlled field from a prescribed set of fields by an amount greater than a given threshold is a minimum. For constructing the irreducible finite dimensional representation of a semisimple idea, we discuss the notion of maximal ideals of the universal enveloping algebra of a semisimple Lie algebra the one-one correspondence between the irreducible representations and maximal ideals. The second part of the book deals with probability theory and we introduce Brownian motion, Poisson process and some of the features assoociated with such processes. The third part of the book covers basic antenna theory including far field radiation pattern by a current source at a given frequency, quantum electrodynamics within a cavity described by the coupling of the second quantized Maxwell and Dirac fields and how to control these cavity fields so as to get a far field radiation pattern having prescribed statistics in a coherent state of cavity photons and Fermions. This portion of the book also introduces the quantum Boltzmann equation in the presence of electromagnetic fields and how from the resulting nonlinear evolution of the density matrix, we can compute the refractive index of materials from quantum averages of electric and magnetic dipole moments. We demonstrate how the refractive index computed in this quantum mechanical way will generally be field dependent. We also show that if gravitational effects are taken into account, for example the metric tensor of space-time in an expanding universe, then Dirac's equation will have to be modified by the presence of background curvature and hence the resulting quantum Boltzmann equation will contain gravitational terms thereby causing the refractive index to depend on the space-time metric of gravitation. We also discuss an interesting notion in electrodynamics namely that of specified the charges in space in terms of the singularities of the electric and magnetic fields. This fact leads to the important conclusion that all charges are in fact generated by the singularities of the electromagnetic field and hence one can postulate that the electron's mass and charge is of electromagnetic origin. In this context, we explain how to compute the corrected electron propagator due to external electromagnetic and gravitational fields by formulating an appropriate differential equation for the electron propagator and hence from this corrected propagator, how to determine the shift in the electron's mass due to electromagnetic and gravitational fields.

Author

Table of Contents

1. Matrix Theory 1–50
2. Probability Theory 51–64
3. Antenna Theory 65–98
4. Miscellaneous Problems 99–108
5. More Problems in Linear Algebra and Functional Analysis 109–184
6. Models for the Refractive Index of Materials and Liquids 185–216
7. More Problems in Probability Theory, Antennas and Refractive Index of Materials 217–248

Detailed Contents

1. Matrix Theory **1–50**

1.1 Perequisites of Linear Algebra 3
1.2 Quotient of a Vector Space 4
1.3 Triangularity of Comuting Operators 5
1.4 Simultaneous Diagonability of a Family of Comuting Normal Operators w.r.t an onb in a Finite Dimensional Complex Inner Product Space 6
1.5 Tensor Products of Vectors and Matrices 7
1.6 The Minimax Variational Principle for Calculating all the Eigenvalues of a Hermitian Matrix 8
1.7 The Basic Decompostition Theorems of Matrix Theory 8
1.8 A Computational Problems in Lie Group Theory 12
1.9 Primary Decompostition Theorem 13
1.10 Existence of Cartan Subalgebra 15
1.11 Exercises in Matrix Theory 24
1.12 Conjugancy Classes of Cartan Subalgebras 28
1.13 Exercises 29
1.14 Appendix: Some Applications of Matrix Theory to Control Theory Problems 30
1.15 Controllability of Supersymmetric Field the Oretic Problems 38
1.16 Controllability of Yang-Mills Gauge Fields in the Quantum Context Using Feynman's Path Integral Approach to Quantum Field Theory 39
1.17 Large Deviations and Control Theory 40
1.18 Approximate Contollability of the Maxwell Equations 42
1.19 Controllability Problems in Quantum Scatering Theory 43
1.20 Kalman's Notion of Controllability and Its Extension to pde's 43
1.21 Controllability in the Context of Representations of Lie Groups 44
1.22 Irreducible Representations and Maximal Ideals 45
1.23 Controllability of the Maxwell-Dirac Equations Using External Classical Current and Field Sources 46

1.24 Controllability of the EEG Signals on the Brain Surface Modeled as a Spherical Surface by Influencing the Infinitesimal Dipoles in the Cells of the Brain Cortex to Vary in Accord to Sensory Perturbations 48
1.25 Control and Relativity 49

2. Probability Theory **51–64**
2.1 The Basic Axioms of Kolmogorov 54
2.2 Exercises 54
2.3 Exercises on Stationary Stochastic Processes, Spectra and Polyspectra 57
2.4 A Research Problem Based on Problem 58
2.5 Exercises on the Construction of the Integral w.r.t a Probability Measure 61
2.6 Exercises on Stationarity, Dynamical Systems and Ergodic Thery 63

3. Antenna Theory **65–98**
3.1 Course Outline 65
3.2 The Far Field Poynting Vector 67
3.3 Exercises 69
3.4 Order of Magnitudes in quantum Antenna Theory 70
3.5 The Notion of a Fermionic Coherent State and its Application to the Computation of the Quantum Statistical Moments of the Quantum Electromagnetic Field Generated by Electrons and Positors Within a Quantum Antenna 73
3.6 Calculating the Moments of the Radiation Field Produced by Electrons and Positrons in the Far Field when the Fermions are in a Coherent State 78
3.7 Controlling the Classical em Fields Interacting with the Dirac Field so that the Mean Value of the em Field Radiated by the Resulting Dirac Second Quantized Current in a Fermionic Coherent State is as Close as Possible to a Given Deterministic Pattern in Space and Simultaneously the Mean Square Fluctuations of this Field in a Fermionic Coherent State are Minimized 79

3.8 Approximate Analysis of a Rectangular Quantum Antenna 81
3.9 Remark on the Perturbation in the Quantum Dirac Field and the Quantum Electromangetic Field Interacting with Each Other Caused by Further Interaction of the Dirac Field with a Classical Control em Field and Interaction of the Quantum Electromagnetic Field with a Control Classical Current 87
3.10 Quantum Antennas Constructed Using Supersymmetric Field Theories 91
3.11 Quantization of the Maxwell and Dirac Field in a Background Curved Metric of Spacetime. 93
3.12 Relationship Between the Electron Self Energy and the Electron Propagator 96
3.13 Electron Self Energy Corrections Induced by Quantum Gravitational Effects 97

4. Miscellaneous Problems 99–108
4.1 A Problem in Robotics 99
4.2 More on Root Space Decompostion of a Semisim-ple Lie Algebra 100
4.3 A Project Proposal for Developing an Ex-perimental Setup for Transmitting Quantum States Over a Channel in the Presence of An Eavesdropper 103
4.4 A Problem in Lie Group Theory 106

5. More Problems in Linear Algebra and Functional Analysis 109–184
5.1 Riesz Representation Theorem 109
5.2 Lie's Theorem on Solvable Lie Algebras 110
5.3 Engel's Theorem on nil-representation of a Lie Algebra 113
5.4 Aperture Antenna Pattern Fluctuations 114
5.5 Spectral Theorem Using Gelfand-Naimark Theorem 116
5.6 The Atiyah-Singer Index Theorem: A supersymmetric Proof 122
5.7 Replicas, Regular Elements, Jordan Decomposition and Cartan Subalgebras 122
5.8 Lecture Plan, Matrix Theory 128
5.9 More Assignment Problems in Probability Theory 131
5.10 Multiple Choice Questions on Probability Theory 133

5.11 Design of a Quantum Unitary Gate Using Superstring Theory with Noise Analysis Based on the Hudson-Parthasarathy Quantum Stochastic Calculus 136
5.12 Study Projects in Probability Theory: Construction of Brownian Motion, Law of the Iterarted Logarithm 137
5.13 Quantum Boltzmann Equation for a Systerm of Particles Interacting with a Quantum Electromagnectic Field 141
5.14 Device Physics in a Semiconductor Using the Classical Boltzmann Transport Equation 143
5.15 Describing the Value of a Point Charge and Its Location in Space in Terms of the Electrostatic Potential Generated by It 143
5.16 Calculating the Masses of *N* Gravitating Particles and Their Postitons and Their Trajectories from Measurement of the Gravitational Potential Distribution in Space-time Using the Newtonian Theory 150
5.17 The Quantum Boltzmann Equation for a Plasma 154
5.18 Some Other Remarks on Lie Algebras 160
5.19 Question Paper on Matrix Theory 164
5.20 Study Project on Quantum Antennas 165
5.21 Heat and Mass Transfer Equations in a Fluid 166
5.22 Quantum Electodynamics in a Background Medium Described by a Permittivity and Permeability Function 167
5.23 Temperature and Field Dependence of Re-fractive Index 171
5.24 Quantum Statistical Field Theory 173
5.25 Root Space Decompositions of the Complex Classical Lie Algebras 176

6. Models for the Refractive Index of Materials and Liquids 185–216
6.1 Quantum Electrodynamics with the Electronic Charge Expressed in Terms of the Quantum Fields 186
6.2 Calculating the Masses of *N* Gravitating Paticles and Their Positions and Their Trajectories from Measurement of the Gravitational Potential Distribution in Space-time Using the Newtonian Theory 193
6.3 The Quantum Boltzmann Equation for a Plasma 196
6.4 Quantum Electrodynamics in a Background Medium Described by a Permittivity and Per-meability Function 202

6.5 Models for the Refractive Index of a Material Based on Classical and Quantum Physics 204
6.6 Quantum Statistical Field Theory 207
6.7 Relating the Refractive Index of a Material to the Metric Tensor of Space-time 209
6.8 Cosmologiccal Effects on the Refractive Index 212

7. More Problems in Probability Theory, Antennas and Refractive Index of Materials 217–248
7.1 Levy's Modulus of Continuity for Brownian Motion 217
7.2 Test 2: Antennas and Wave Propagation 220
7.3 Article Submitted to the Quantum Information Processing Journals for Publication 222

Chapter 1

Matrix Theory

Prerequisites of linear algebra.

Fields, rings, vector spaces over a field, modules over a ring, algebras, ideals in a ring and an algebra, bases for vector spaces, linear transformations in a vector space, basis for a vector space, matrix of a linear transformation relative to a basis, inner product spaces, unitary, Hermitian and normal operators in an inner product space, spectral theorem for normal operators.

[1] Quotient of a vector space by another space.

[2] Simultaneous triangulability of commuting matrices relative to an onb.

[3] Simultaneous diagonability of commuting normal matrices relative to an onb.

[4] Tensor products of vector spaces.

[5] Variational principles for calculating the eigenvalues of a Hermitian matrix.

[6] Positive definite matrices.

[7] The basic decomposition theorems of matrix theory.

[a] Row reduced Echelon form.

[b] Spectral theorem for normal matrices.

[c] Polar decomposition.

[d] Singular value decomposition.

[e] QR decomposition based on the Gram-Schmidt orthonormalization process.

[f] LDU decomposition of positive definite matrices.

[8] Applications of matrix theory to finite state quantum systems.

[a] Schrodinger and Heisenberg evolution in finite dimensional Hilbert spaces.

[b] Different kinds of unitary gates for finite state quantum computation: CNOT, Swap, Fredkin, Toffoli, phase gate, Quantum Fourier transform gate.

[c] Perturbation theory for quantum systems in finite dimensional state space.

[8] Lie groups and Lie algebras.

[a] Group action on a differentiable manifold.

[b] Differential of a group action on a manifold.

[c] Differentiable and analytic structures on a group: The notion of a Lie group based on differentiability of the composition and inversion operation.

[d] The Lie algebra of vector fields on a differentiable manifold

[e] The notion of Lie algebra of a Lie group determined by left invariance of vector fields.

[f1] Identification of the Lie algebra of a Lie group with its tangent space at the identity.

[f2] The exponential map from the Lie algebra into a Lie group.

[f3] Examples when the exponential map is not surjective: $O(3)$

[f4] The one-one correspondence between Lie algebra elements and one parameter subgroups of a Lie group.

[g] Structure constants associated with a basis for a Lie algebra.

[h] Examples of linear Lie groups and their Lie algebras: The classical Lie groups

[i] Examples of nonlinear Lie groups taken from dynamical systems.

[j] Homotopy and covering groups of a Lie group.

[k] The universal enveloping algebra of a Lie algebra.

[l] The invariant bilinear form on a Lie algebra.

[m] Solvable and semisimple Lie algebras.

[n] Simple Lie algebras, root space decomposition and Cartan's classification theory based on the theory of roots.

[9] Representation theory for Lie groups and Lie algebras.

[a] Completely reducible representations.

[b] irreducible representations.

[c] Classification of finite dimensional irreducible representations of a semisimple Lie algebra based on dominant integral weights: The Cartan-Weyl-HarishChandra theory.

[d] The character of a representation.

[e] Representations of compact groups:Schur Lemmas and the Peter-Weyl theory.

[f] Characters of a compact semisimple group: Weyl's character formula.

[g] Applications of representation theory to image processing problems.

[10] Algebraic varieties in a multivariable polynomial ring.

[11] Prime ideals and maximal ideals of a ring.

[12] Grothendieck's generalization of an algebraic variety to schemes on a ring consisting of prime ideals.

[13] Some remarks on algebraic groups.

[a] Examples.

[b] Flag variety.

[c] The Grassmannian variety and the Schubert variety.

[d1] The relationship between irreducible representations of an algebraic group and Flag varieties.

[d2] The relationship between Schubert cells in a Grassmannian variety and the Schubert variety constructed from the Bruhat decomposition of a semisimple algebraic group associated with a parabolic subgroup.

[e] Plucker's coordinates on a Grassmannian variety.

[f] Quadratic relations satisfied by Plucker's coordinates on the Grassmannian variety.

1.1 Perequisites of linear algebra

[1] A **field** is a set of elements $\mathbb{F}$ that has two binary operations called addition and multiplication, a zero element 0 and a unit element 1 such that under addition, $\mathbb{F}$ is an Abelian group with identitity 0, under multiplication, $\mathbb{F} - \{0\}$ (ie, $\mathbb{F}$ without the zero element) is an Abelian group with identity 1 and multiplication distributes over addition. Examples of fields are

[a] $\mathbb{R}, \mathbb{C}, \mathbb{Q}$, the set of real, complex and rational numbers.

[b] $\{0, 1, ..., p-1\}$ where p is any prime, with addition and multiplication defined as for real numbers but modulo p.

[c] If x is any indeterminate, and $\mathbb{F}$ is any field, then $\mathbb{Q}_F[x]$, the set of all rational functions in x over $\mathbb{F}$, ie, the set of all ratios of polynomials in x with the coefficients of the polynomials coming from $\mathbb{F}$ and with the denominator polynomial never being the zero polynomial is a field.

[2] A **vectorspace** over a field $\mathbb{F}$ is a set of elements endowed with a binary operation called vector addition and denoted by $+$, a scalar multiplication, ie, a map $\mathbb{F} \times V \to V$ denoted by a dot and a zero element called the zero vector so that under vector addition, V forms an Abelian group with identity being the zero vector $\mathbf{0}$, scalar multiplication distributes over vector addition, if $c_1, c_2 \in \mathbb{F}, \mathbf{x} \in V$, then

$$c_1.(c_2.\mathbf{x}) = (c_1c_2).\mathbf{x},$$

$$(c_1 + c_2).\mathbf{x} = c_1.\mathbf{x} + c_2.\mathbf{x}$$

if 0 denotes the zero element of $\mathbb{F}$, then

$$0.\mathbf{x} = \mathbf{0},$$

and if 1 denotes the unit element in $\mathbb{F}$, then

$$1.\mathbf{x} = \mathbf{x}$$

if $-\mathbf{x}$ denotes the additive inverse of $\mathbf{x}$ in the Abelian group V (under vector addition), then,

$$-1.\mathbf{x} = -\mathbf{x}$$

Examples of vector spaces:

[a] $V = \mathbb{F}^n$, the set of all column vectors (or equivalently row vectors), ie, ordered n-tuples with the n entries in $\mathbb{F}$ and vector addition and scalar multiplication being defined component wise, ie,

$$[c_1, ..., c_n]^T + [d_1, ..., d_n]^T = [c_1 + d_1, ..., c_n + d_n]^T, c_i, d_i \in \mathbb{F}, i = 1, 2, ..., n$$

$$c.[c_1, ..., c_n]^T = [cc_1, ..., cc_n]^T, c, c_1, ..., c_n \in \mathbb{F}$$

In other words, vector addition and scalar multiplication in $\mathbb{F}^n$ are induced by addition and multiplication in $\mathbb{F}$.

3. A **ring** is a set R with two binary operations $+, .$ and a zero element 0 such that under $+$, R is an Abelian group and under ., R is a semigroup, (A semigroup has all the properties of the group like closure under composition, associativity under composition except that there may not be any identity and inverse of an element) and such that . distributes over $+$. Clearly, a field is a special case of a ring. Some examples of a rings that is not fields are

[a] $\mathbb{F}[x]$, the set of all polynomials in an indeterminate x with coefficients taken from the field $\mathbb{F}$. This is called the ring of polynomials over the field $\mathbb{F}$. Note that a polynomial will not have a multiplicative inverse unless it is a constant polynomial. $\mathbb{F}[x]$ is an example of a commutative/Abelian ring.

[b] $R = M_n(\mathbb{F})$, the space of $n \times n$ matrices with elements from $\mathbb{F}$. The . and $+$ operations in this ring are respectively matrix multiplication and matrix addition. Note that this ring is non-Abelian.

[c] As a generalization of $[a]$, we define R to be the commutative ring of all functions f from a set X to a field $\mathbb{F}$. The . operation is simply the ordinary multiplication of functions with the multiplication operation induced by that in $\mathbb{F}$ and the addition operation is likewise the ordinary addition operation of functions induced by addition in $\mathbb{F}$.

[d] As a generalization of $[a]$ and $[c]$ we consider the ring R of all polynomials in an indeterminate x with coefficients coming from $M_n(\mathbb{F})$, ie, all $n \times n$ matrix polynomials.

1.2 Quotient of a vector space

2. Simultaneous triangulability of a family of commuting matrices relative to an o.n.b.

3. Simultaneous diagonability of a family of commuting normal matrices relative to an o.n.b.

2[a]. First we prove the triangulability of a matrix over the complex field relative to an o.n.b.

Let V be a complex inner product space of dimension n and let $T : V \to V$ be a linear operator. T has at least on eigenvalue: Say

$$Te_1 = c_1 e_1, c_1 \in \mathbb{C}, \| e_1 \| = 1$$

Define

$$W_1 = N(T - c_1 I)$$

Then

$$W_1 \neq \{0\}$$

Consider the quotient vector space

$$V/W_1`, dim(V/W_1) < n$$

W_1 is T-invariant and hence

$$T_1 : V/W_1 \to V/W_1, T(x + W_1) = T(x) + W_1$$

is a well defined linear operator. We define an inner product on V/W_1 as follows: For $x, y \in V$,

$$< x + W_1, y + W_2 >=< P^\perp x, P^\perp y >$$

where P is the orthogonal projection of V onto W and $P^\perp = I - P$. It is easy to see that this is a well defined inner product on V/W_1. Now by the induction hypothesis on the vector space dimension, T_1 can be brought into upper triangular form relative to an onb say $\{f_k + W_1 : k = 1, 2, ..., m\}$ of V/W_1. We note that

$$P^\perp f_k + W_1 = f_k + W_1$$

and hence we can replace the above onb for V/W_1 by $\{P^\perp f_k + W_1 : k = 1, 2, ..., m\}$. Then by construction, $P^\perp f_k, k = 1, 2, ..., m$ is an orthonormal set in V and further the span of this set is orthogonal to W_1. Thus, if we choose an orthogonormal basis $\{e_1, ..., e_{n-m}\}$ for W_1, then it is immediate that T is upper-triangular w.r.t the onb

$$\mathcal{B} = \{P^\perp f_1, ..., P^\perp f_m, e_1, ..., e_{n-m}\}$$

for V.

1.3 Triangulability of comuting operators

Now we prove simultaneous triangulability of a family of commuting operators on a finite dimensional complex inner product space w.r.t an onb. Let $dim_{\mathbb{C}} V = n < \infty$ and let $\mathcal{F}$ be a family of commuting operators in this space. Choose one operator say T from this family and choose an eigenvalue c_1 with a corresponding normalized eigenvector e_1 for this operator:

$$Te_1 = c_1 e_1, \| e_1 \|= 1$$

Then

$$W_1 = N(T - c_1)$$

is a non-zero subspace of V and is invariant under every element of $\mathcal{F}$ since all these elements commute with T. Thus for each $S \in \mathcal{F}$, we can define a linear operator

$$S_1 : V/W_1 \to V/W_1$$

by

$$S_1(x + W_1) = S(x) + W_1, x \in V$$

and it is immediate that $\{S_1 : S \in \mathcal{F}\}$ is a commuting family of operators in V/W_1 and hence by the induction hypothesis on the vector space dimension, this family can be simultaneously triangulated w.r.t an onb for V/W_1. Let $\{f_k + W_1 : k = 1, 2, ..., m\}$ be such an onb. Then again by the induction hypothesis, $\mathcal{F}$ restricted to W_1 can also be simultaneously triangulated w.r.t an onb for it. Denote this onb by $\{e_1, ..., e_{n-m}\}$. It is immediate then that $\mathcal{F}$ is simultaneously triangulable relative to the onb $\{P^{\perp}f_1, ..., P^{\perp}f_m, e_1, ..., e_{n-m}\}$ for V. The proof is complete.

Reference: This problem was taken from
Rajendra Bhatia, "Matrix Analysis", Springer.

1.4 Simultaneous diagonability of a family of commuting normal operators w.r.t an onb in a finite dimensional complex inner product space

Let $\mathcal{F}$ be a commuting family of normal operators in V with $dim_{\mathbb{C}}V = n < \infty$. We first choose an element $T \in \mathcal{F}$ and a vector e_1 of unit norm such that

$$Te_1 = c_1e_1$$

for some $c_1 \in \mathbb{C}$. Again define

$$W_1 = N(T - c_1)$$

Choose any $S \in \mathcal{F}$. Then, W_1 is S-invariant since $[T, S] = 0$. Now define $S_1 : V/W_1 \to V/W_1$ in the usual way. We claim that S_1 is normal. Indeed, let $u, v \in V$. Then,

$$< u + W_1|S_1^*S_1|v + W_1 >=< S(u) + W_1|S(v) + W_1 >=< P^{\perp}Su|P^{\perp}Sv >$$

where P is the orthogonal projection onto W_1. Then, P is expressible as a function of T using the spectral theorem for normal operators, ie, $P = f(T)$. From that it is immediate that S commutes with P and hence also with $P^{\perp} = I - P$. Since $P^{\perp}$ is Hermitian, S^* also commutes with $P^{\perp}$. Then, using this and the normality of S,

$$< P^{\perp}Su|P^{\perp}Sv >=< u|S^*P^{\{perp}Sv >=$$

$$< u|P^{\perp}S^*S|v >=< u|P^{\perp}SS^*|v >$$

$$=< [S^*]_1(u + W_1)|[S^*]_1(v + W_1 >$$

On the other hand, we claim that

$$[S^*]_1 = S_1^*$$

for the following reason:

$$< u + W_1|[S^*]_1|v + W_1 >=< u + W_1|S^*(v) + W_1 >=< P^\perp u|P^\perp S^* v >$$

$$=< u|P^\perp S^* v >$$

while on the other hand,

$$< u + W_1|S_1^*|v + W_1 >=< S_1(u + W_1)|v + W_1 >=< S(u) + W_1|v + W_1 >$$

$$=< P^\perp S(u)|P^\perp v >=< Su|P^\perp v >=< u|S^* P^\perp v >$$

$$=< u|P^\perp S^* v >$$

proving the claim. Thus, we get

$$< u+W_1|S_1^* S_1|v+W_1 >=< S_1^*(u+W_1)|S_1^*(v+W_1) >=< u+W_1|S_1 S_1^*|v+W_1 >$$

and therefore

$$S_1^* S_1 = S_1 S_1^*$$

ie, S_1 is normal for all $S \in \mathcal{F}$. Moreover, it is clear that $\{S_1 : S \in \mathcal{F}\}$ is a commuting family since if $S, L \in \mathcal{F}$, then

$$S_1 L_1(u+W_1) = S_1(L(u)+W_1) = SL(u)+W_1 = LS(u)+W_1 = L_1 S_1(u+W_1), u \in V$$

Thus, $\{S_1 : S \in \mathcal{F}\}$ is a commuting family of normal operators in V/W_1 and by the induction hypothesis (on the dimension of the vector space), it follows that this family is simultaneously diagonable w.r.t an onb say $\{f_k + W_1 : k = 1, 2, ..., m\}$ for V/W_1. Likewise by the same induction hypothesis, $\mathcal{F}$ is simultaneously diagonable on W_1 w.r.t. an onb say $\{e_1, .., e_{n-m}\}$ for W_1. In other words, for any $S \in \mathcal{F}$, we have

$$S(f_k) - c_k(S) f_k \in W_1, k = 1, 2, ..., m$$

Thus,

$$S(P^\perp f_k) - c_k(S) P^\perp f_k = P^\perp(S f_k - c_k(S) f_k) = 0$$

This proves that $\{P^\perp f_k, k = 1, 2, ..., m, e_1, ..., e_{n-m}\}$ is an onb for V that simultaneously diagonalizes $\mathcal{F}$.

1.5 Tensor products of vectors and matrices

[a] Tensor products of vector spaces.

[b] The symmetric tensor product and permanents.

[c] The antisymmetric tensor product and determinants.

[d] Tensor product of matrices.

[e] Eigenvalues and eigenvectors of the tensor product, antisymmetric tensor product and symmetric tensor product of matrices.

1.6 The minimax variational principle for calculating all the eigenvalues of a Hermitian matrix

Let X be an $n \times n$ Hermitian matrix and let $\{v_1, ..., v_n\}$ be an orthonormal eigenbasis for X with corresponding eigenvalues $\{c_1, ..., c_n\}$ so that $c_1 \geq c_2 \geq ... \geq c_n$. Let M be any k dimensional subspace of $\mathbb{C}^n$. Consider the $n-k+1$ dimensional subspace $W_k = span\{v_k, ..., v_n\}$ of $\mathbb{C}^n$. It is clear that

$$dim(M \cap W_k) = dimM + dimW_k - dim(M + W_k) \geq k + (n-k+1) - n = 1$$

and hence we can choose

$$v \in M \cap W_k, \| v \|= 1$$

and therefore,

$$< v|X|v > \leq c_k$$

It follows that

$$inf_{x \in M} < x|X|x > \leq c_k$$

Thus,

$$sup_{dimN=k} inf_{x \in N} < x|X|x > \leq c_k$$

Now choosing

$$N = span\{v_1, ..., v_k\}$$

it follows that $dimN = k$ and $v_k \in N$,

$$< v_k|X|v_k >= c_k$$

Thus we have proved

$$sup_{dimN=k} inf\{x \in N\} < x|X|x >= c_k$$

This is the first minimax theorem for computing the eigenvalues of a Hermitian matrix. Now we prove the second minimax theorem.

1.7 The basic decomposition theorems of matrix theory

[a] LDU decomposition of a positive definite matrix. Let R be a positive definite matrix. It can be looked upon as the correlation matrix of a random vector $\mathbf{X}$. We write

$$\mathbf{X} = [X_1, ..., X_n]^T$$

and then Gram-Schmidt orthonormalize this vector relative to the standard inner product on $L^2(\Omega, \mathcal{F}, P)$: $< u, v >= \mathbb{E}(uv)$. Then we get an orthonormal set $\{e_1, e_2, ..., e_n\}$ of random variables in $L^2(\Omega, \mathcal{F}, P)$ where

$$X_1 = a(1,1)e_1, X_2 = a(2,1)e_1 + a(2,2)e_2, ..., X_k = a(k,1)e_1 + ... + a(k,k)e_k, k = 1, 2, ..., n$$

writing

$$\mathbf{e} = [e_1, ..., e_n]^T$$

we can express this as

$$\mathbf{X} = \mathbf{Le}$$

where $\mathbf{L}$ is the lower triangular matrix $((a(i,j)))$. Then taking correlations on both sides gives us

$$\mathbf{R} = \mathbb{E}(\mathbf{XX}^T) = \mathbf{L}\mathbb{E}(\mathbf{ee}^T)\mathbf{L}^T = \mathbf{LL}^T$$

which is known as the LU decomposition of $\mathbf{R}$. Defining $\mathbf{B}$ as $\mathbf{L}$ except that its diagonal entries are all 1 and defining the diagonal matrix

$$\mathbf{D} = diag[a(i,i)^2, i = 1, 2, ..., n]$$

we get

$$\mathbf{R} = \mathbf{BDB}^T$$

with B lower triangular and having ones on its diagonal. This is called the LDU decomposition of the matrix $\mathbf{R}$. This decomposition is important in linear prediction theory of stochastic processes as its derivation suggests.

There is another way to derive the LDU decomposition of a positive definite matrix R. Using the spectral decomposition theorem, we can write $R = XX^T$ where X is a square matrix. Then using the QR decomposition, write

$$X^T = QY$$

where Y is upper triangular and Q is orthogonal. This can be achieved for example, by applying the Gram-Schmidt decomposition to the columns of X^T or equivalently to the rows of X. In other words

$$Q = [e_1, e_2, ..., e_n], e_a^T e_b = \delta_{ab}$$

$$X^T = [x_1, x_2, ..., x_n]$$

Then,

$$x_1 = y(1,1)e_1, x_2 = y(1,2)e_1 + y(2,2)e_2, ..., x_k = y(1,k)e_1 + ... y(k,k)e_k, k = 1, 2, ..., n$$

Then,

$$R = XX^T = Y^TQ^TQY = Y^TY$$

and Y^T is lower triangular.

[b] The QR decomposition of any rectangular matrix. First assume that $A \in \mathbb{C}^{m \times n}$. Let

$$A = [a_1, ..., a_n], a_j \in \mathbb{C}^{m \times 1}, j = 1, 2, ..., n$$

Choose a permutation matrix P so that if

$$B = AP$$

so that the columns of B are obtained by permuting the columns of A, then

$$B = [b_1, ..., b_n]$$

and $b_1, ..., b_r$ are linearly independent while $b_{r+1}, ..., b_n$ can be expressed as linear combinations of $b_1, ..., b_r$. We can thus write

$$[b_{r+1}, ..., b_n] = [b_1, .., b_r]C$$

for some appropriate $r \times n$ matrix C. Then

$$B = [B_1|B_1C] = B_1[I_r|C], B_1 = [b_1, ..., b_r] \in \mathbb{C}^{m \times r}$$

Apply the Gram-Schmidt process to the columns of B_1 which are linearly independent and then obtain

$$B_1 = [e_1, ..., e_r]R_1$$

where $e_1, ..., e_r$ are orthonormal and R_1 is an $r \times r$ upper triangular matrix. Extend the set $\{e_1, ..., e_r\}$ to an orthonormal basis $\{e_1, ..., e_m\}$ for $\mathbb{C}^m$ and thus define the unitary matrix

$$Q = [e_1, ..., e_m] \in \mathbb{C}^{m \times m}$$

Define

$$R_2 = \begin{pmatrix} R_1 \\ 0 \end{pmatrix} \in \mathbb{C}^{m \times r}$$

Then, we can write

$$B_1 = QR_2$$

and

$$B = [QR_2|QR_2C] = QR_2[I|C]$$

and hence

$$A = BP^{-1} = QR_2[I|C]P^{-1}$$

In the special case when the columns of A are all linearly independent, this formula reduces to

$$A = QR_2$$

An application of the QR decomposition to linear least squares problems.

Let A be an $m \times n$ matrix of full column rank n. Then, by Gram-Schmidt orthornormalization of its columns, we get

$$A = Q_1 R_1$$

where Q_1 is $m \times n$ with orthonormal columns and R_1 is $n \times n$ upper triangular and non-singular, ie, with nonzero columns on its diagonals. Extend Q_1 by appending more orthonormal columns to Q so that Q becomes a $m \times m$ orthogonal matrix and define

$$R = \begin{pmatrix} R_1 \\ 0 \end{pmatrix} \in \mathbb{C}^{m \times n}$$

Then

$$A = QR$$

Now consider the problem of minimizing

$$E(\theta) = \| x - A\theta \|^2, \theta \in \mathbb{R}^n$$

We have

$$\| x - QR\theta \|^2 = \| y - R\theta \|^2, y = Q^T x$$

Write

$$\theta = \begin{pmatrix} \theta_1 \\ \theta_2 \end{pmatrix}$$

where

$$\theta_1 \in \mathbb{R}^n, \theta \in \mathbb{R}^{m-n}$$

Then, clearly,

$$E(\theta) = \| y_1 - R_1\theta_1 \|^2 + \| y_2 \|^2$$

where

$$y = \begin{pmatrix} y_1 \\ y_2 \end{pmatrix}, y_1 \in \mathbb{R}^n, y_2 \in \mathbb{R}^{m-n}$$

Thus, $E(\theta)$ is minimized when and only when

$$\theta_1 = R_1^{-1} y_1$$

In other words, the set of all θ' that minimize $E(\theta)$ are of the form

$$\theta = \begin{pmatrix} R_1^{-1} y_1 \\ \theta_2 \end{pmatrix}, \theta_2 \in \mathbb{R}^{m-n}$$

1.8 A computational problem in Lie group theory

Let G be a Lie group and $\mathfrak{g}$ is Lie algebra. For $X \in \mathfrak{g}$, define the left invariant vector field on G by the formula

$$v_X f(g) = \frac{d}{dt} f(g.exp(tX))|_{t=0}$$

where $f : G \to \mathbb{C}$ is a differentiable function.

Now do the following:

[a] Take $G = SL(2, \mathbb{R})$ so that $\mathfrak{g}$ has a basis H, X, Y where

$$H = \begin{pmatrix} 1 & 0 \\ 0 & -1 \end{pmatrix},$$

$$X = \begin{pmatrix} 0 & 1 \\ 0 & 0 \end{pmatrix}, Y = \begin{pmatrix} 0 & 0 \\ 1 & 0 \end{pmatrix}$$

Express v_H, v_X, v_Y as linear first order differential operators in the coordinates (t, x, y) where $g \in G$ is parametrized as

$$g = g(t, x, y) = exp(tH + xX + yY)$$

[b] Consider the Lie group $G = SO(3)$. Its Lie algebra is the real vector space spanned by all 3×3 skew symmetric real matrices. We choose the standard basis for this Lie algebra, namely

$$X_1 = \begin{pmatrix} 0 & 0 & 0 \\ 0 & 0 & -1 \\ 0 & 1 & 0 \end{pmatrix}$$

$$X_2 = \begin{pmatrix} 0 & 0 & 1 \\ 0 & 0 & 0 \\ -1 & 0 & 0 \end{pmatrix}$$

$$X_3 = \begin{pmatrix} 0 & -1 & 0 \\ 1 & 0 & 0 \\ 0 & 0 & 0 \end{pmatrix}$$

Prove that the element

$$R(n) = exp(n_1 X_1 + n_2 X_2 + n_3 X_3), n_1, n_2, n_3 \in \mathbb{R}$$

defines a rotation around the axis

$$\hat{n} = (n_1, n_2, n_3)/\sqrt{n_1^2 + n_2^2 + n_3^2}$$

by an angle

$$\phi = \sqrt{n_1^2 + n_2^2 + n_3^2}$$

in the counterclockwise sense. Evaluate the vector fields $v_{X_k}, k = 1, 2, 3$ in terms of the Euler angles which parametrize $R \in SO(3)$ in terms of the Euler angles:

$$R = R_z(\phi)R_x(\theta)R_z(\psi) = R(\phi, \theta, \psi)$$

by expressing

$$\frac{d}{dt}f(R(\phi, \theta, \psi)exp(tX_k))|_{t=0}, k = 1, 2, 3$$

as a linear partial differential operators of the first order in (ϕ, θ, ψ). Now writing

$$v_{X_k} = v_{k1}(\phi, \theta, \psi)\partial/\partial\phi + v_{k2}(\phi, \theta, \psi)\partial/\partial\theta + v_{k3}\partial/\partial\psi, k = 1, 2, 3$$

compute

$$f(\phi, \theta, \psi) = det((v_{ij}(\phi, \theta, \psi)))^{-1}$$

and prove using general arguments that

$$f(\phi, \theta, \psi)d\phi.d\theta, d\psi$$

is the Haar measure on $SO(3)$, ie, invariant under left and right translations.

1.9 Primary decomposition theorem

Let T be a linear operator in a finite dimensional complex vector space V. Prove the primary decomposition theorem:

$$V = \bigoplus_{k=1}^{r} W_k$$

where

$$W_k = N((T - c_k)^{m_k}), k = 1, 2, ..., r$$

with $c_1, ..., c_r$ being the distinct eigenvalues of T and

$$p(t) = \Pi_{k=1}^{r}(t - c_k)^{m_k}$$

being the minimal polynomial of T. Prove that

$$W_k = \bigcup_{m \geq 1} N((T - c_k)^m)$$

Let E_k denote the projection onto W_k corresponding to this decomposition, ie,

$$R(E_k) = W_k, E_kE_j = 0, k \neq j, \sum_{k=1}^{r} E_k = I$$

Show that an operator S in V commutes with T iff

$$E_kSE_m = 0\forall k \neq m$$

iff

$$S = \sum_{k=1}^{m} E_kSE_k$$

Now suppose $ad(T)(S) = [T,S]$ leaves all the subspaces $W_k, k = 1,2,...,r$ invariant. Then, we claim that S also shares this same property. Indeed, consider for some $k \neq l$ the operator $S_{kl} = E_kSE_l$. Then since the $E_j's$ commute with T (in fact, they are polynomials in T), it follows that

$$E_k ad(T)(S)E_l = E_k[T,S]E_l = [T, S_{kl}]$$

and the lhs is zero by hypothesis. Thus, S_{kl} commutes with T and hence leaves all the subspaces $W_j, j = 1,2,...,r$ invariant. This means that $S_{kl} = 0$ and since $k \neq l$ are arbitary, it follows that $S = \sum_{j=1}^{r} E_jSE_j$ leaves all the subspaces $W_j, j = 1,2,...,r$ invariant. Now we can prove the following theorem: If for some positive integer n, $ad(T)^n(S) = 0$, then S leaves all the subspaces $W_j, j = 1,2,...,r$ invariant. In fact, $0 = ad(T)^n(S) = [T, ad(T)^{n-1}(S)]$ implies that $ad(T)^{n-1}(S)$ leaves all the subspaces $\{W_j\}$ invariant and by induction, it follows that $ad(T)^j(S), j = n-2, n-3, ..., 1, 0$ also leave all these subspaces invariant. The proof of the theorem is complete.

Now we wish to use the primary decomposition theorem to prove that

$$V = [\bigcup_{m\geq 1} N(T^m)] \cap [\bigcap_{m\geq 1} R(T^m)]$$

Indeed, suppose T has no zero eigenvalue. Then, the primary decomposition theorem for T reads

$$V = \bigoplus_{k=1}^{r} N((T-c_k)_k^m)$$

where none of the $c_k's$ are zero. Now, if for some k

$$v \in N((T-c_k)^{m_k})$$

then

$$(T-c_k)^{m_k}v = 0$$

and this equation can be expressed as

$$(-1)^{m_k}c_kv + Tf_k(T)v = 0$$

where $f(T)$ is a polynomial in T. Since $c_k \neq 0$, it follows that

$$v = Tg_k(T)v$$

where g_k is also a polynomial. By induction, it follows that

$$v \in R(T^m), m = 1, 2, ...$$

and hence

$$v \in \bigcap_{m \geq 1} R(T^m)$$

Since T has no zero eigenvalue, we also have

$$N(T^m) = 0, m = 1, 2, ...$$

and the proof of the theorem for this case is complete. Now suppose that zero is an eigenvalue of T. Then we can take $c_1 = 0$ and since $c_k \neq 0, k > 1$, the result follows in the same way from the primary decomposition theorem by noting that

$$W_1 = N(T^{m_1}) = \bigcup_{m \geq 1} N(T^m),$$

and for $k > 1$,

$$W_k = N((T - c_k)^{m_k}) \subset \bigcap_{m \geq 1} R(T^m)$$

Remark: If $T^m v = 0$ for some positive integer m and simultaneously $v \in R(T^r)$ for all $r \geq 1$, then we can write $v = T^r v_r, r \geq 1$ and hence $T^{m+r} v_r = 0$ for all $r \geq 1$ and hence $v_r \in N(T^{m_1})$ for all $r \geq 1$. Therefore, $v = T^{m_1+r} v_{m_1+r} = 0$. This proves that

$$[\bigcup_{m \geq 1} N(T^m)] \bigcap [\bigcap_{m \geq 1} R(T^m)] = \{0\}$$

1.10 Existence of Cartan subalgebra

Now we apply this result to the proof of the existence of a Cartan subalgebra of a semisimple Lie algebra leading thereby to the root space decomposition and hence Cartan's classification of all the simple Lie algebras.

Let $\mathfrak{g}$ be a any finite dimensional Lie algebra over the complex field. A Lie subalgebra $\mathfrak{h}$ of $\mathfrak{g}$ is said to be a Cartan algebra (or a Cartan subalgebra of $\mathfrak{g}$), if (a) $\mathfrak{h}$ is nilpotent and (b) $\mathfrak{h}$ is its own normalizer in $\mathfrak{g}$. ie $X \in \mathfrak{g}$ and $[X, \mathfrak{h}] \subset \mathfrak{h}$ together imply $X \in \mathfrak{h}$. We first show that a Cartan algebra is maximal nilpotent. Indeed, suppose that $\mathfrak{n}$ is a nilpotent Lie algebra that properly contains $\mathfrak{h}$. Consider the Lie algebra $\mathfrak{n}/\mathfrak{h}$ with its canonical Lie bracket. This is a non-trivial Lie algebra and it is nilpotent. Thus, its adjoint representation is a nil representation. By a basic theorem in nil-representations, it follows therefore that there exists a non-zero element $\xi = X + \mathfrak{h} \in \mathfrak{n}/\mathfrak{h}$ (ie $X \notin \mathfrak{h}$) such that

$ad(\mathfrak{n}/\mathfrak{h})(\xi) = \mathfrak{h}$ (Note that $\mathfrak{h}$ is the zero element in $\mathfrak{n}/\mathfrak{h}$. It follows therefore that $[\mathfrak{n}, X] \subset \mathfrak{h}$ and this obviously implies $[X, \mathfrak{h}] \subset \mathfrak{h}$. This contradicts the fact that a Cartan algebra is its own normalizer.

Regular element: Now given a $X \in \mathfrak{g}$ define the characteristic polynomial of $ad(X)$:

$$p_X(t) = det(tI - ad(X)) = t^n + t^{n-1}c_1(X) + ... + tc_{n-1}(X) + c_n(X)$$

and set

$$l(X) = min(m : c_{n-m}(X) \neq 0)$$

Then $0 \leq l(X) \leq n$. Define

$$rk(\mathfrak{g}) = min(l(X) :\in \mathfrak{g})$$

Further, if $\mathfrak{h}$ is any subspace of $\mathfrak{g}$, define

$$\zeta_{\mathfrak{h}}(X) = det(ad(X)|_{\mathfrak{g}/\mathfrak{h}})$$

If $\mathfrak{h}$ is a subspace of $\mathfrak{g}$ such that $rk(\mathfrak{g}) = dim\mathfrak{h}$ and if X a regular element of $\mathfrak{h}$ in $\mathfrak{g}$, then it is clear that $\zeta_{\mathfrak{h}}(X) \neq 0$ since in this case, $l(X) = rk(\mathfrak{g}) = dim\mathfrak{h}$.

Construction of Cartan algebras: Let X be any regular element in $\mathfrak{g}$. Define

$$\mathfrak{h} = \mathfrak{h}_X = \bigcup_{m>0} N(ad(X)^m)$$

Then, we claim that $\mathfrak{h}$ is a Cartan algebra. First observe that $\mathfrak{h}$ is closed under linear combinations and under the Lie bracket operation. Indeed, $Y, Z \in \mathfrak{h}$ imply

$$ad(X)^m([Y, Z]) = \sum_{r=0}^{m} \binom{m}{r} [ad(X)^r(Y), ad(X)^{m-r}(Z)] = 0$$

for sufficiently large m. Here we have used that $ad(X)$ is a derivation on $\mathfrak{g}$ and $ad(X)^r(Y) = 0$ or $ad(X)^{m-r}(Y) = 0$ for all sufficiently large m and all $r = 0, 1, ..., m$. Next we show that $\mathfrak{h}$ is a nilpotent algebra. Define $\mathfrak{h}'$ to be the set of all $Y \in \mathfrak{h}$ for which $\zeta_{\mathfrak{h}}(Y) \neq 0$. Clearly we have the direct sum decomposition

$$\mathfrak{g} = \bigcup_{m \geq 1} N(ad(X)^m) \oplus \bigcap_{m \geq 1} R(ad(X)^m) =$$

$$\mathfrak{h} \oplus q$$

Now $Y \in \mathfrak{h}$ implies that for some positive integer m, $ad(X)^m(Y) = 0$ and this implies that

$$(ad(adX))^m(ad(Y)) = 0$$

which implies that $ad(Y)$ leaves $\mathfrak{h} = \bigcup_{m\geq 1} N(ad(X)^m)$ invariant. Thus, $\mathfrak{h}$ is a Lie subalgebra of $\mathfrak{g}$. This also implies that $ad(Y)$ leaves $q = \bigcup_{m\geq 1} R(ad(X)^m)$ invariant as discussed above in the context of consequences of the primary decomposition theorem. $ad(X)$ is nilpotent on $\mathfrak{h}$ and non-singular on q. Thus, since X has been assumed to be a regular element, it follows that $rk(\mathfrak{g}) = dim\mathfrak{h}$. Now suppose that $Y \in \mathfrak{h}'$. Then, $ad(Y)$ is non-singular on $\mathfrak{g}/\mathfrak{h}$ and hence also non-singular on q. Thus, $l(Y) \leq l(X) = rk(\mathfrak{g}) \leq l(Y)$. Thus, $l(Y) = rk(\mathfrak{g}) = dim\mathfrak{h}$. Since $ad(Y)$ is non-singular on q and since $ad(Y)$ leaves $\mathfrak{h}$ invariant, it must necessarily follow that if $ad(Y)^m(Z) = 0$ for some positive integer m and $Z \in \mathfrak{g}$, then $Z \in \mathfrak{h}$. In other words, we have shown that $\mathfrak{h}_Y \subset \mathfrak{h}$ and since

$$dim\mathfrak{h}_Y = l(Y) = dim\mathfrak{h}$$

it follows that

$$\mathfrak{h}_Y = \mathfrak{h}$$

It follows in particular that $ad(Y)$ is nilpotent on $\mathfrak{h}$ (since by definition, it is nilpotent on $\mathfrak{h}_Y$). Since $\mathfrak{h}'$ is a non-empty open subset of $\mathfrak{h}$ and since Y is an arbitrary element of $\mathfrak{h}'$, it follows that $ad(Y)$ is nilpotent for every $Y \in \mathfrak{h}$. This, completes the proof that $\mathfrak{h} = \mathfrak{h}_X$ is a Cartan algebra for every regular $X \in \mathfrak{g}$.

Now let $\mathfrak{h}$ be a Cartan subalgebra of $\mathfrak{g}$. Let $\mathfrak{h}'$ denote the set of regular elements in $\mathfrak{h}$. Let $X \in \mathfrak{h}'$. We claim that $\mathfrak{h} = \mathfrak{h}_X$ where $\mathfrak{h}_X$ has been defined above as $\bigcup_{n\geq 1} N(ad(X)^n)$. Indeed, since $\mathfrak{h}$ is nilpotent, it follows that $ad(X)$ is nilpotent on $\mathfrak{h}$ and therefore $\mathfrak{h} \subset \mathfrak{h}_X$. On the other hand, we have seen above that $\mathfrak{h}_X$ is a Cartan algebra and hence maximally nilpotent. Since $\mathfrak{h}$ is also a Cartan algebra, it is also maximally nilpotent. Hence $\mathfrak{h} = \mathfrak{h}_X$. In other words, we have proved that every Cartan algebra is of the form $\mathfrak{h}_X$ for some regular element X and in fact or argument shows that $\mathfrak{h} = \mathfrak{h}_X$ for any regular element X of $\mathfrak{h}$.

Now let $\mathfrak{g}$ be any semisimple Lie algebra. Let $\mathfrak{h}$ be any Cartan algebra. Then, since $\mathfrak{h}$ is a nilpotent Lie algebra and hence also a solvable Lie algebra and $H \to ad(H)$ is a representation of $\mathfrak{h}$, in $\mathfrak{g}$, it follows that there is a basis for $\mathfrak{g}$ relative to which all the operators $ad(H), H \in \mathfrak{h}$ are upper-triangular (not necessarily upper triangular). Now, if $N\mathfrak{h}$ is such that $ad(N)$ is nilpotent on $\mathfrak{g}$, then its matrix relative to this basis will be strictly upper-triangular and hence

$$< N, H >= Tr(ad(N).ad(H)) = 0, H \in \mathfrak{h}$$

If we are able therefore to prove that $< .,. >$ is non-singular on $\mathfrak{h} \times \mathfrak{h}$, then it would follow from the above relation that $N = 0$, ie, $\mathfrak{h}$ does not have any nilpotent elements. To prove this claim, we choose $X \in \mathfrak{h}'$. Then, $\mathfrak{h} = \mathfrak{h}_X$. Let $X = S + N_1$ be the Jordan decomposition of X into its semisimple and nilpotent components. Then, from the basic property of the Jordan decomposition, $[X, S] = 0$ and hence $S \in \mathfrak{h}_X = \mathfrak{h}$. further since $ad(X) = ad(S) + ad(N)$ and $ad(S)$ is semisimple while $ad(N)$ is nilpotent, it follows that $ad(X)$ and $ad(S)$

have the same characteristic polynomial and hence $S \in \mathfrak{h}'$. Hence, $\mathfrak{h} = \mathfrak{h}_S$. Since $ad(S)$ is semisimple, it then follows that for any $Y \in \mathfrak{h}$, we have $ad(S)(Y) = 0$, ie, $[S, Y] = 0$ and in fact, $\mathfrak{h} = N(ad(S))$. let $q = [S, \mathfrak{g}]$=R(ad(S)). Since $ad(S)$ is semisimple, we get

$$\mathfrak{g} = \mathfrak{h} \oplus q$$

Then,

$$< [Y, S], H >=< Y, [S, H] >= 0, H \in \mathfrak{h}, Y \in \mathfrak{g}$$

In other words,

$$< q, \mathfrak{h} >= 0$$

and hence, from the non-singularity of $< .,. >$ on $\mathfrak{g}$ (Cartan's criterion for semisimplicity of $\mathfrak{g}$), it must follow that $< .,. >$ is non-singular on $\mathfrak{h}$. This completes the proof of the claim.

Next, we show that if $\mathfrak{g}$ is semi-simple, then any Cartan algebra $\mathfrak{h}$ is maximal Abelian. Let $X \in \mathfrak{h}'$ and let $X = S + N$ be its Jordan decomposition. Then, $ad(X)(N) = [X, N] = 0$ and hence $N \in \mathfrak{h}_X = \mathfrak{h}$. Since $ad(N)$ is also nilpotent, by what we proved above, $N = 0$. Hence $X = S$ is semisimple. Thus $ad(X)^m(Y) = 0$ for some $Y, m > 0$ implies $[X, Y] = 0$. In other words, we have that $\mathfrak{h}'$ is Abelian and since $ad(X)$ is semisimple for all $X \in \mathfrak{h}'$, it follows that $ad(\mathfrak{h}')$ can be simultaneously diagonalized in $\mathfrak{g}$. By taking limits noting that $\mathfrak{h}'$ is a non-empty open subset of $\mathfrak{h}$ and hence dense in it it follows immediately that $ad(\mathfrak{h})$ is simultaneously diagonable and hence Abelian. From the faithfulness of the adjoint representation of a semisimple Lie algebra, it then follows that $\mathfrak{h}$ is also Abelian and since it is maximally nilpotent, it is also necessarily maximal Abelian. (Suppose $[\mathfrak{h}, Y] = 0$. Then, $ad(X)(Y) = 0$ for $X \in \mathfrak{h}'$ and hence $Y \in \mathfrak{h}_X = \mathfrak{h}$). Thus, we have proved the following fundamental result in the theory of semisimple Lie algebras:

Theorem: if $\mathfrak{g}$ is a semisimple Lie algebra then there exists a Lie subalgebra $\mathfrak{h}$ of $\mathfrak{g}$ such that $\mathfrak{h}$ is its own normalizer in $\mathfrak{g}$ and secondly, $\mathfrak{h}$ is Maximal Abelian, ie, $\mathfrak{h}$ is Abelian and $Y \in \mathfrak{g}, [\mathfrak{h}, Y] = 0$ implies $Y \in \mathfrak{h}$. $\mathfrak{h}$ is called a Cartan subalgebra of $\mathfrak{g}$.

Later on we shall prove that if the field is complex, $\mathfrak{g}$ has exactly one Cartan subalgebra upto conjugacy, ie, if $\mathfrak{h}_1, \mathfrak{h}_2$ are any two Cartan subalgebras of $\mathfrak{g}$, then there exists a $g \in G$ such that $\mathfrak{h}_2 = Ad(g).\mathfrak{h}_1$.

We have already shown assuming that $\mathfrak{g}$ is a semisimple Lie algebra, that if $\mathfrak{h}$ is a Cartan subalgebra, then there exists a regular element $X \in \mathfrak{g}$ such that $\mathfrak{h} = \mathfrak{h}_X$. Now we prove a slightly stronger version of this result namely: There exists a finite number $X_1, ..., X_r$ of regular elements in $\mathfrak{g}$ such that if $\mathfrak{h}$ is any Cartan algebra, then there exists an $i \in \{1, 2, ..., r\}$ and an $x \in G$ such that $\mathfrak{h} = \mathfrak{h}_i^x$ where $\mathfrak{h}_i = \mathfrak{h}_{X_i}$ and further that $\mathfrak{h}_i, i = 1, 2, ..., r$ are all mutually non-conjugate.

Note: Since we have made use of the Jordan decomposition on a semsimple Lie algebra in our construction of a Cartan subalgebra and proofs of some of these properties, we shall give a proof of this theorem in what follows.

Let $\mathfrak{g}$ be a semisimple Lie algebra and let $X \in \mathfrak{g}$. Write $ad(X) = T + U$ where U is nilpotent on $\mathfrak{g}$ and T is semisimple on $\mathfrak{g}$ and $[T, U] = 0$. This is the Jordan decomposition of a linear operator in a vector space. We claim that T and U are also derivations on $\mathfrak{g}$. In fact, we known from Chevalley's theory of replicas that T and U are replicas of $ad(X)$ and therefore since $ad(X)$ is a derivation, so are T and U.

Remark: Let V be a vector space and $\beta : V \times V \to V$ bilinear and let D be a derivation on V w.r.t β, ie,

$$D\beta(X, Y) = \beta(DX, Y) + \beta(X, DY), X, Y \in V$$

Then let L be a replica of D. It is easy to see that L is then also a derivation on V w.r.t β. Indeed, consider the mapping $\eta : V \otimes V^* \otimes V^* \to B(V \times V, V)$ where $B(V \times V, V)$ is the space of V valued bilinear forms on V. The map η is defined by

$$\eta(X \otimes f_1 \otimes f_2)(U, V) = f_1(U)f_2(V)X$$

and then extending η bilinearly w.r.t its first two arguments. It is then easy to see that η is a vector space isomorphism. Note that if we choose a basis $\{e_1, ..., e_n\}$ for V, and if $\{e_1^*, ..., e_n^*\}$ denotes the corresponding dual basis, then

$$\beta(U, V) = \sum_{i,j} \beta(e_i, e_j)e_i^*(U)e_j^*(V)$$

$$= \eta(\sum_{i,j} \beta(e_i, e_j) \otimes e_i^* \otimes e_j^*)(U, V)$$

or equivalently,

$$\beta = \eta(\sum_{i,j} \beta(e_i, e_j) \otimes e_i^* \otimes e_j^*)$$

which proves that η is surjective and therefore also injective since the dimension of $V \otimes V^* \otimes V^*$ equals $(dimV)^3$ which is also the dimension of the space of all V valued bilinear forms on V. Writing

$$\beta_1(U, V) = \beta(DU, V), \beta_2(U, V) = \beta(U, DV)$$

We have

$$\beta_1(U, V) = \sum_{ij} \beta_1(e_i, e_j)e_i^*(U)e_j^*(V)$$

$$= \sum_{i,j} \beta(De_i, e_j)e_i^*(U)e_j^*(V)$$

$$= \sum_{i,j,k} [D]_{ki}\beta(e_k, e_j)e_i^*(U)e_j^*(V)$$

$$= \sum_{j,k} \beta(e_k, e_j)(D^T e_k^*)(U)e_j^*(V)$$

and therefore,

$$\beta_1 = \eta(\sum_{i,j} \beta(e_i, e_j) \otimes (D^T e_i^*) \otimes e_j)$$

Likewise,

$$\beta_2 = \eta(\sum_{i,j} \beta(e_i, e_j) \otimes e_i^* \otimes D^T e_j^*)$$

Further,

$$D(\beta(U,V)) = \sum_{i,j} (D\beta(e_i, e_j)) e_i^*(U) e_j^*(V)$$

$$= \eta(\sum_{i,j} (D\beta(e_i, e_j)) \otimes e_i^* \otimes e_j^*)(U,V)$$

and hence the derivation property of D, namely

$$D(\beta(U,V)) = \beta_1(U,V) + \beta_2(U,V)$$

can equivalently be expressed as

$$\sum_{i,j} [(D\beta(e_i, e_j)) \otimes e_i^* \otimes e_j^* - \beta(e_i, e_j) \otimes D^T e_i^* \otimes e_j^* - \beta(e_i, e_j) \otimes e_i^* \otimes D^T e_j^*] = 0$$

which in the notation of replicas means that

$$D_{1,2}(\sum_{i,j,k} \beta(e_i, e_j) \otimes e_i^* \otimes e_j^*) = 0 - - - (1)$$

Now let D be a derivation on V w.r.t β and let

$$D = T + U$$

be the Jordan decomposition of D. Then, it is known that T and U are also replicas of D and in particular (1) implies that

$$T_{1,2}(\sum_{i,j,k} \beta(e_i, e_j) \otimes e_i^* \otimes e_j^*) = 0,$$

$$U_{1,2}(\sum_{i,j,k} \beta(e_i, e_j) \otimes e_i^* \otimes e_j^*) = 0$$

ie T and S are also derivations on V w.r.t β. Now assuming that $\mathfrak{g}$ is a semisimple Lie algebra, we observe that ad is faithful, ie, injective on $\mathfrak{g}$. Indeed, this follows from the fact that $ad(X) = 0$ for some non-zero $X \in \mathfrak{g}$ would imply that $[X, \mathfrak{g}] = 0$ with some non-zero X and this would imply that $\mathfrak{g}$ has a non-zero centre which contradicts the semisimplicity of $\mathfrak{g}$. Thus, writing the Jordan decomposition of the derivation $ad(X)$ as

$$ad(X) = T + U$$

we get that the semisimple and nilpotent components of $ad(X)$, namely T, U are also derivations and hence are inner, ie, there exist $S, N \in \mathfrak{g}$ such that $T = ad(S), U = ad(N)$. Then, from the faithfulness of ad, it follows that

$$X = S + N$$

with $ad(S)$ semisimple $ad(N)$ nilpotent and $[ad(S), ad(N)] = 0$ or equivalently, $ad([S, N]) = 0$ or equivalently, $[S, N] = 0$. This is the celebrated Jordan decomposition of a semisimple Lie algebra.

Remark: Let D be a derivation on $\mathfrak{g}$ where $\mathfrak{g}$ is a semisimple Lie algebra. Then, D is inner. Indeed, for $X, Y \in \mathfrak{g}$, we have

$$ad(DX)(Y) = [DX, Y] = D[X, Y] - [X, DY] = D[X, Y] - ad(X)(DY)$$

$$= Doad(X)(Y) - ad(X)(DY)$$

or equivalently,

$$ad(DX) = [D, ad(X)], \forall X \in \mathfrak{g}$$

Now, by non-degeneracy of the Cartan-Killing form $< ., . >$ on a semisimple Lie algebra, we have a unique $X \in \mathfrak{g}$ such that

$$Tr(D.ad(Y)) =< X, Y >= Tr(ad(X).ad(Y))\forall Y \in \mathfrak{g}$$

since $Y \to Tr(D.ad(Y))$ is a linear functional on $\mathfrak{g}$. This shows that

$$Tr(D'.ad(Y)) = 0\forall Y \in \mathfrak{g}$$

where

$$D' = D - ad(X)$$

Now, D' is also a derivation since D and $ad(X)$ and hence an application of the above formula to D' in place of D, we get

$$ad(D'Y) = [D', ad(Y)], Y \in \mathfrak{g}$$

and hence

$$< D'Y, Z >= Tr(ad(D'Y)ad(Z)) = Tr([D', ad(Y)].ad(Z))$$

Now,

$$Tr([D', ad(Y)]ad(Z)) = Tr(D'ad(Y)ad(Z)) - Tr(ad(Y)D'ad(Z)) =$$

$$Tr(D'(ad(Y)ad(Z) - ad(Z)ad(Y)) = Tr(D'[ad(Y), a(Z)]) =$$

$$Tr(D'ad([Y, Z])) = 0$$

by what we just proved. Therefore,

$$< D'Y, Z >= 0\forall Y, Z \in \mathfrak{g}$$

and hence by non-degeneracy of $< ., . >$, it follows that

$$D'Y = 0 \forall Y \in \mathfrak{g}$$

and hence

$$D' = 0$$

ie,

$$D = ad(X)$$

This completes the proof that D is inner. This result has been used in the proof of the Jordan decomposition on a semisimple Lie algebra.

Remark: Let T be a linear operator in a complex vector space V. Let T' denote the commutant of T, ie, T' is the set of all operators in V that commute with T. Let T'' denote the double commutant of T, ie, the commutant of T', ie, the set of all operators that commute with every operator in T'. Then, it is easy to show that T'' is precisely the set of all polynomials in T with complex coefficients. In fact, this can be proved by restricting T' and T'' to the space $W_k = N((T - c_k)^{m_k}) = R(E_k)$ where

$$I = E_1 + ... + E_r$$

or equivalently,

$$V = W_1 \oplus ... \oplus W_r$$

is the primary decomposition of T with

$$p(t) = \Pi_{k=1}^{r}(t - c_k)^{m_k}$$

being the minimal polynomial of T. Note that W_k is also T' and T''-invariant since $T'' \subset T'$. Then by restricting to W_k the claim reduces to proving that if $X \in (cI + N)''$ where c is a complex scalar and N nilpotent in V, then X is a polynomial $cI + N$. Further reduction of this problem can be achieved by using the Jordan decomposition of N. In other words, proving the claim reduces to proving that if J_c is a Jordan matrix in V, ie,

$$J_c = cI + Z$$

where $c \in \mathbb{C}$ and Z has ones on the first superdiagonal and all the other entries as zero, then J_c'' is precisely the set of all polynomials in J_c.

Remark: Let T be an operator in V. we claim that if S a replica of T, then S is a polynomial in T with the constant term in the polynomial being zero, ie, $S = Tf(T)$ where f is a polynomial. To see this, we first define an isomorphism $\mu : V \otimes V^* \to L(V)$, where $L(V)$ is the space of all linear operators on V by

$$\mu(v \otimes w^*)(x) = w^*(x)v, x, v \in V, w^* \in V^*$$

and then extending μ by bilinearity. Then,

$$ad(\mu(v \otimes w^*))(T)(x) = [\mu(v \otimes w^*), T](x)$$

$$= w^*(Tx)v - w^*(x)Tv = (T^T w^*)(x)v - w^*(x)Tv = \mu(v \otimes T^T w^*)(x) - \mu(Tv \otimes w^*)(x)$$

$$= -\mu(T_{1,1}(v \otimes w^*))(x)$$

Equivalently,

$$ad(T)(\mu(v \otimes w^*))(x) = \mu(T_{1,1}(v \otimes w^*))(x)$$

or equivalently,

$$ad(T)o\mu = \mu o T_{1,1}$$

or equivalently,

$$ad(T) = \mu o T_{1,1} o \mu^{-1}, T_{1,1} = \mu^{-1} o ad(T) o \mu$$

If S is a replica of T, then $T_{r,s}\xi = 0$ implies $S_{r,s}\xi = 0$ for any $r, s \geq 0, r + s \geq 1$. Thus, in particular, $T_{1,1}\xi = 0$ implies $S_{1,1}\xi = 0$, or equivalently, in view of the above discussion $ad(T)(U) = 0$ implies $\mu o T_{1,1} o \mu^{-1}(U) = 0$ for an operator U in V implies $T_{1,1} o \mu^{-1}(U) = 0$ implies $S_{1,1} o \mu^{-1}(U) = 0$ implies $\mu o S_{1,1} o \mu^{-1}(U) = 0$ implies $ad(S)(U) = 0$. In other words, S commutes with any operator U that commutes with T, ie, $S \in T''$ and hence by the previous remark, S is a polynomial in T, say $S = f(T)$ where f is a polynomial. Further since S is a replica of T $T\xi = 0$ implies $S\xi = 0$ implies $f(T)\xi = 0$ implies that if c is the constant term in $f(t)$, then $c\xi = 0$. In other words, if T has a zero eigenvalue, then $c = 0$, ie, $S = Tg(T)$ where g is a polynomial. Suppose that zero is not an eigenvalue of T. Then, T is invertible and hence the minimal polynomial of T has the form

$$p(t) = \Pi_{k=1}^{r}(t - c_k)^{m_k}, c_k \neq 0 \forall k$$

Thus,

$$p(t) = c + tq(t), c \neq 0$$

with q a polynomial Then $p(T) = 0$ implies

$$cI + Tq(T) = 0$$

and therefore,

$$T^{-1} = -c^{-1}q(T)$$

ie, T^{-1} is also a polynomial in T. It then follows that if S is a replica of T, then

$$S = f(T) = c_0 I + Tg(T) = c_0 TT^{-1} + Tg(T) = T((-c_0/c)q(T) + g(T))$$

ie, S can be expressed as a polynomial in T where the polynomial has zero constant term. We shall now prove the following important result: S is a replica of T iff for all $r, s \geq 0, r + s \geq 1$, we have that $S_{r,s} = p_{r,s}(T_{r,s})$ where $p_{r,s}$ is a polynomial with zero constant term.

1.11 Exercises in Matrix Theory

:

[1] Let X be a square matrix of size $n \times n$. Prove that X has a polar decomposition:

$$X = UP$$

where U is unitary and P is positive semidefinite. In case X is non-singular, show that

$$P = \sqrt{X^*X}, U = X(X^*X)^{-1/2}$$

where $\sqrt{Q}$ denotes the unique positive semidefinite square root of a positive semidefinite matrix Q.

hint: Show that if Q is positive definite of size $n \times n$, then Q can have atmost 2^n distinct square roots out of which exactly one is positive semidefinite. Show that X, X^*X and $|X| = \sqrt{X^*X}$ all have the same nullspace and hence the same nullity and hence the same rank and hence $R(|X|)^\perp|$ and $R(X)^\perp$ also have the same dimension. Show that the operator $U_1 : R(|X|) \to R(X)$ defined by $U_1|X|x = Xx \forall x \in \mathbb{C}^n$ is a well defined unitary operator. Do this by showing that the lengths of $|X|x$ and Xx are the same. Hence show that there exists a unitary operator U_2 from $R(|X|)^\perp \to R(X)^\perp$. Let $V = \mathbb{C}^n$ and hence

$$V = R(|X|) \oplus R(|X|)^\perp = R(X) \oplus R(X)^\perp$$

Define a linear operator $U : V \to V$ by the relation that U restricted to $R(|X|)$ equals U_1 and U restricted to $R(|X|)^\perp$ equals U_2. Show that U is unitary and

$$X = U_1|X| = U|X|$$

[2] Deduce the singular value decomposition from the polar decomposition: If X is a matrix of size $m \times n$ of rank r (Note that $r \leq min(m,n)$), then there exist unitary matrices $U \in \mathbb{C}^{m\times m}, V \in \mathbb{C}^{n\times n}$ and a matrix $D \in \mathbb{C}^{m\times n}$ having the block structure

$$D = \begin{pmatrix} D_1 & 0 \\ 0 & 0 \end{pmatrix}$$

where

$$D_1 = diag[\sigma_1, ..., \sigma_r], \sigma_1, ..\sigma_r > 0$$

such that

$$X = UDV^*$$

[3] Prove the Riesz representation theorem in an infinite dimensional Hilbert space $\mathcal{H}$: If $f : \mathcal{H} \to \mathbb{C}$ is a bounded linear functional, ie $\| f \| = sup_{x\neq 0}|f(x)|/\| x \|$, then there exists a unique vector $z_f \in \mathcal{H}$ such that

$$f(x) = < z_f, x >, x \in \mathcal{H}$$

hint: Assume $\mathcal{H}$ to be separable which guarantees the existence of an orthonormal basis $\{e_n : n \geq 1\}$. Then show that for any $x \in \mathcal{H}$, we have

$$x = \sum_n e_n < e_n, x >$$

Deduce using the linearity and boundedness of f that

$$f(x) = \sum_n f(e_n) < e_n, x >$$

Finally, show that

$$\infty >\| f \|^2 = \sum_n |f(e_n)|^2$$

and hence that

$$z_f = \sum_n \bar{f}(e_n) e_n \in \mathcal{H}$$

is well defined.

[4] If $T : V_1 \to V_2$ is a linear transformation from one vector space V_1 to another vector space V_2, both assumed to be finite dimensional, then letting V_1^*, V_2^* denote the vector space of linear functionals on V_1 and V_2 respectively, define the transpose T' of T as a transformation $T' : V_2^* \to V_1^*$ by

$$T'f(x) = f(Tx), x \in V_1, f \in V_2^*$$

Show that T' is a well defined linear transformation and if we have a third vector space V_3 and a linear transformation $S : V_2 \to V_3$, then

$$(ST)' = T'S'$$

For these statements to be true, do we actually require V_1, V_2, V_3 to be finite dimensional or can we drop this condition ?

[5] This problem gives us some properties of the derivative of a function with values in a Hilbert space.

Let $\mathcal{H}$ be an infinite dimensional Hilbert space and $x : \mathbb{R} \to \mathcal{H}$ be a function such that

$$lim_{\delta \to 0}(x(t+\delta) - x(t))/\delta = y(t) \in \mathcal{H}$$

exists for $t \in (-a, +a)$ where the convergence is in the sense of the norm induced by the inner product in $\mathcal{H}$. Deduce that if $z \in \mathcal{H}$ is arbitrary, then

$$\frac{d}{dt} < x(t), z >=< y(t), z >, t \in (-a, a), z \in \mathcal{H}$$

where now the convergence is in the usual sense on the real line. We shall write $dx(t)/dt = y(t), t \in (-a, a)$ and say that $x(t)$ is differentiable in $(-a, a)$ with derivative $dx(t)/dt$ equal to $y(t)$ for $t \in (-a, a)$. Now prove that if $x_1(t)$ and

$x_2(t)$ assume values in $\mathcal{H}$ and are differentiable in $(-a, a)$, then deduce that for any $c_1, c_2 \in \mathbb{C}$, $c_1x_1(t) + c_2x_2(t)$ is also differentiable in $(-a, a)$ with derivative given by

$$\frac{d}{dt}(c_1x_1(t) + c_2x_2(t)) = c_1\frac{dx_1(t)}{dt} + c_2\frac{dx_2(t)}{dt}, t \in (-a, a)$$

hint: Prove using the definitions

$$lim_{\delta\to 0} \parallel \frac{x_k(t+\delta) - x_k(t)}{\delta} - dx_k(t)/dt \parallel = 0, k = 1, 2$$

and the triangle inequality that

$$lim_{\delta\to 0} \parallel \frac{c_1x_1(t+\delta) + c_2x_2(t+\delta)}{\delta} - (c_1dx_1(t)/dt + c_2dx_2(t)/t) \parallel = 0$$

[6] Let T be a linear operator on a finite dimensional complex vector space V with minimal polynomial $p(t) = \Pi_{k=1}^{r}(t - c_k)^{m_k}$, $c_k's$ distinct and $m_k > 0$. The primary decomposition of T is

$$V = \bigoplus_{k=1}^{r} W_k, I = \sum_{k=1}^{r} E_k, R(E_k) = W_k, E_kE_j = 0, k \neq j$$

Then if S is another linear operator in V such that $[T, S]$ leaves each of the W_k invariant, then show that S also shares the same property.

hint: Note that the $E_k's$ are all polynomials in T and hence commute with T. Also note that an operator L leaves each of the W_k invariant iff

$$L = \sum_{k,j} E_kSE_j = \sum_k E_kSE_k$$

or equivalently, iff

$$E_kLE_j = 0\forall k \neq j$$

Hence, for all $k \neq j$,

$$0 = E_k[T, S]E_j = [T, E_kSE_j]$$

and hence E_kSE_j leaves every W_i invariant and in particular, W_j invariant. This means that $E_kSE_j = 0$ for all $k \neq j$ and hence

$$S = \sum_{k,j} E_kSE_j = \sum_k E_kSE_k$$

proving that S leaves every W_k invariant. Note that in the proof, we have used the easily verified fact that if $[T, U] = 0$, then U leaves every W_k invariant because

$$UE_k = E_kU, W_k = R(E_k)$$

Note that U commutes with E_k because E_k is a polynomial in T.

[7] Let G be a connected Lie group and let H be a connected Lie subgroup of G. Thus if $\mathfrak{g}$ and $\mathfrak{h}$ are respectively the Lie algebras of G and H, then the respective exponential maps are surjective, ie, $exp(\mathfrak{g}) = G$ and $exp(\mathfrak{h}) = H$. Show then that if $N(H)$ denotes the normalizer of H in G, ie, the set of all $g \in G$ for which $gHg^{-1} \subset H$, and if $\mathfrak{n}(H)$ denotes the Lie algebra of $N(H)$, then

$$\mathfrak{n}(H) = \{X \in \mathfrak{g} : [X, \mathfrak{h}] \subset \mathfrak{h}\}$$

[8] This problem discusses a method for obtaining all the finite dimensional irreducible representations of the Lie group $SL(2, \mathbb{C})$ or equivalently of its Lie algebra $sl(2, \mathbb{C})$. Let $G = SL(2, \mathbb{C})$, ie, the set of all 2×2 complex matrices having determinant one. Let $sl(2, \mathbb{C}) = \mathfrak{g}$, the Lie algebra of G. Show that $sl(2, \mathbb{C})$ is the set of all 2×2 complex matrices having trace zero. Show that a basis for $sl(2, \mathbb{C})$ is given by $\{H, X, Y\}$, where

$$H = \begin{pmatrix} 1 & 0 \\ 0 & -1 \end{pmatrix},$$

$$X = \begin{pmatrix} 0 & 1 \\ 0 & 0 \end{pmatrix},$$

$$Y = \begin{pmatrix} 0 & 0 \\ 1 & 0 \end{pmatrix}$$

Note prove the following identities:

[a]

$$[H, X] = 2X, [X, Y] = -2Y, [X, Y] = H$$

and hence if π is any representation of $sl(2, \mathbb{C})$ in a vector space V, then

$$[\pi(H), \pi(X)] = 2\pi(X), [\pi(H), \pi(Y)] = -2\pi(Y), [\pi(X), \pi(Y)] = \pi(H)$$

Let now π be in particular a representation of $sl(2, \mathbb{C})$ in a finite dimensional complex vector space V. Show that if v is a vector in V such that

$$\pi(H)v = \lambda v$$

for some $\lambda \in \mathbb{C}$, then

$$\pi(H)\pi(X)v = (\lambda + 2)\pi(X)v, \pi(H)\pi(Y)v = (\lambda - 2)\pi(Y)v$$

Hence, deduce from the finite dimensionality of V that there exist a nonzero vector $v_0 \in V$ and a $\lambda_0 \in \mathbb{C}$ such that

$$\pi(X)v_0 = 0, \pi(H)v_0 = \lambda_0 v_0$$

and that there exists a smallest positive integer l such that $\pi(Y)^l v_0 = 0$. By smallest, we mean that $\pi(Y)^{l-1}v_0 \neq 0$. Deduce that $V_0 = \{\pi(Y)^m v_0 : 0 \leq m \leq l-1\}$ is an invariant subspace for π in V, ie,

$$\pi(sl(2,\mathbb{C}))(V_0) \subset V_0$$

and hence deduce that if π is assumed to be irreducible, then $\{\pi(Y)^m v_0 : 0 \leq m \leq l-1\}$ is a basis for V. Now prove that if n is any positive integer, then

$$[\pi(H),\pi(X)^n] = 2n.\pi(X)^n, [\pi(H),\pi(Y)] = -2n\pi(Y)^n$$

and

$$[\pi(X),\pi(Y)^n] =$$

$$\pi(H).\pi(Y)^{n-1} + \pi(Y)\pi(H).\pi(Y)^{n-2} + ... + \pi(Y)^{n-1}\pi(H)$$

$$= [-2(n-1)\pi(Y)^{n-1} + \pi(Y)^{n-1}\pi(H)] + [-2(n-2)\pi(Y)^{n-1} + \pi(Y)^{n-1}\pi(H)]$$

$$+ ... + [0.\pi(Y)^{n-1} + \pi(Y)^{n-1}\pi(H)]$$

$$= -2(0+1+2+...+(n-1)]\pi(Y)^{n-1} + n\pi(Y)^{n-1}\pi(H)$$

$$= -2n(n-1)\pi(Y)^{n-1} + n\pi(Y)^{n-1}\pi(H)$$

1.12 Conjugacy classes of Cartan subalgebras

Consider a semisimple Lie algebra $\mathfrak{g}$ and let $\mathfrak{g}'$ denote the set of all of its regular elements. Let $\mathfrak{g}_i, i = 1,2,...,r$ denote all the connected components of $\mathfrak{g}$. For each $i = 1,2,...,r$, choose an $X_i \in \mathfrak{g}_i$. Then since X_i is regular, it follows that $\mathfrak{h}_i = \mathfrak{h}_{X_i}$ is a Cartan algebra for each i. Let $\mathfrak{h}$ be any Cartan algebra. Then $\mathfrak{h} = \mathfrak{h}_X$ for some regular X as we have already seen above. Since X is regular, it follows that $X \in \mathfrak{g}_i$ for some i. Our aim is to show that $\mathfrak{h} = \mathfrak{h}_X$ is conjugate to $\mathfrak{h}_i$. We would then have established that any semisimple Lie algebra (finite dimensional) has a finite set of non-conjugate Cartan subalgebras such that any Cartan subalgebra is conjugate to one in this set. Let $\mathfrak{h}'_X$ denote the set of regular elements in $\mathfrak{h}_X$, ie, $\mathfrak{h}'_X = \mathfrak{h}_X \cap \mathfrak{g}'$. Let $\mathfrak{h}_{X+}$ denote the connected component of $\mathfrak{h}'_X$ that contains X and define $\mathfrak{b}_X = (\mathfrak{h}_{X+})^G$. Then, it is clear that $\mathfrak{b}_X$ is connected. Choose any $Z \in \mathfrak{b}_X$. Then, Z is regular and hence the Cartan algebra $\mathfrak{h}_Z$ defined. We claim that $\mathfrak{b}_Z = \mathfrak{b}_X$. Indeed, Z^y is a regular element in $\mathfrak{h}_X$ for some $y \in G$ and hence, $\mathfrak{h}_Z^y = \mathfrak{h}_{Z^y} = \mathfrak{h}_X$ which implies that $\mathfrak{h}_Z^{'y} = \mathfrak{h}'_X$ and therefore

$$\mathfrak{b}_Z = (\mathfrak{h}'_Z)^G = (\mathfrak{h}'_X)^G = \mathfrak{b}_X$$

proving the claim. Now let $U,V \in \mathfrak{g}_i$. We claim that either $\mathfrak{b}_U = \mathfrak{b}_V$ or else $\mathfrak{b}_U \cap \mathfrak{b}_V = \phi$. Indeed, suppose $Z \in \mathfrak{b}_U \cap \mathfrak{b}_V$. Then, $Z \in \mathfrak{b}_U$ which implies as shown above that $\mathfrak{b}_Z = \mathfrak{b}_U$ and likewise, $\mathfrak{b}_Z = \mathfrak{b}_V$. Thus, $\mathfrak{b}_U = \mathfrak{b}_V$, thereby proving the claim. Further, $\mathfrak{b}_U$ is an open connected set of regular elements containing U and hence $\mathfrak{b}_U \subset \mathfrak{g}_i$. In other words, we have proved that $\{\mathfrak{b}_U : U \in \mathfrak{g}_i\}$ is a family of connected open sets, each of which is contained in $\mathfrak{g}_i$ and two elements in this family are either disjoint or the same. Further, the union of this whole family is precisely $\mathfrak{g}_i$ since if $U \in \mathfrak{g}_i$, we have that $U \in \mathfrak{b}_U$. It follows from the connectedness of $\mathfrak{g}_i$ that $\mathfrak{b}_U = \mathfrak{g}_i \forall U \in \mathfrak{g}_i$. We have thus shown that $\mathfrak{b}_{X_i} = \mathfrak{g}_i, i = 1, 2, ..., r$. Now, let $\mathfrak{h} = \mathfrak{h}_X$ be as above with $X \in \mathfrak{g}_i$ (Recall that any Cartan subalgebra is of this form for some regular X, and some i). Then we have established that $\mathfrak{b}_X = \mathfrak{b}_{X_i}$. In other words, we have established that $\mathfrak{h}^G_{X+} = \mathfrak{h}^G_{X_i+}$ and this implies that $X = Z^y$ for some $y \in G$ and some $Z \in \mathfrak{h}_{X_i+}$ (Recall that $X \in \mathfrak{h}_{X+}$). Then, $\mathfrak{h} = \mathfrak{h}_X = \mathfrak{h}_{Z^y} = \mathfrak{h}^y_Z = \mathfrak{h}^y_{X_i}$ (since Z, X_i are both regular elements in $\mathfrak{h}_{X_i}$). Therefore, $\mathfrak{h} = \mathfrak{h}^y_i$, ie, $\mathfrak{h}$ is conjugate to $\mathfrak{h}_i$. This completely proves our aim.

Remark: If the underlying field of $\mathfrak{g}$ is complex, then $\mathfrak{g}$ has just one Cartan subalgebra upto conjugacy, ie, any two of its Cartan subalgebras are conjugate.

1.13 Exercises

[1] Let $\mathfrak{h}$ be a Cartan subalgebra of any finite dimensional Lie algebra $\mathfrak{g}$, ie $\mathfrak{h}$ is a Lie subalgebra, nilpotent and its own normalizer. Show that there exists an $X \in \mathfrak{g}'$ ($\mathfrak{g}'$ is the set of regular elements in $\mathfrak{g}$) such that $\mathfrak{h}_X = \mathfrak{h}$ where

$$\mathfrak{h}_X = \bigcup_{m>0} N(ad(X)^m)$$

[2] With $\mathfrak{h}$ any Cartan subalgebra of a Lie algebra $\mathfrak{g}$, show that $X \in \mathfrak{h}$ is regular iff

$$\zeta(X) = det(ad(X)|_{\mathfrak{g}/\mathfrak{h}}) \neq 0$$

hint: $\zeta(X)$ is non-zero for $X \in \mathfrak{h}$ iff $ad(X)|_{\mathfrak{g}/\mathfrak{h}}$ has a zero eigenvalue iff when we write

$$det(tI - ad(X)) = c_k t^k + c_{k+1} t^{k+1} + ... + t^n, n = dim\mathfrak{g}$$

with $c_k \neq 0$, then $k > dim(\mathfrak{h}) = rk(\mathfrak{g})$. Do this by noting that for $X \in \mathfrak{h}$, all the eigenvalues of $ad(X)|_{\mathfrak{h}}$ are zero since by definition, $ad(X)$ is nilpotent on $\mathfrak{h}$ and therefore,

$$det(tI - ad(X)|_{\mathfrak{h}}) = t^l, l = dim(\mathfrak{h}) = rk(\mathfrak{g})$$

[2] With $\mathfrak{h}_X$ as in the problems [1,2] for X regular (ie, $X \in \mathfrak{g}', \mathfrak{h}_X = \bigcup_{m>0} N(ad(X)^m)$), show without assuming that $\mathfrak{h}_X$ is a Cartan algebra that $ad(X)$ is non-singular on $\mathfrak{g}/\mathfrak{h}_X$ and hence $dim(\mathfrak{h}_X) = rk(\mathfrak{g})$(Note that to prove that $ad(X)$ is non-singular on $\mathfrak{g}/\mathfrak{h}_X$, you need not even assume that X is regular since by the definition of $\mathfrak{h}_X$, $ad(X)|_{\mathfrak{g}/\mathfrak{h}_X}$ is non-singular). Now show that with X regular, if $Y \in \mathfrak{h}_X$, then $ad(X)^m(Y) = 0$ for some $m > 0$ and hence $ad(Y)$ leaves $\mathfrak{h}_X$ invariant. Show further that if $\zeta(Y) = det(ad(Y)|_{\mathfrak{g}/\mathfrak{h}_X}) \neq 0$, then Y is regular and hence $\mathfrak{h}_Y = \mathfrak{h}_X$. Deduce from this that for such a Y, $ad(Y)$ is nilpotent on $\mathfrak{h}_X$. Then, observe that if $Y \in \mathfrak{h}_X$ is arbitrary, we can write $Y = limY_n$ with $\zeta(Y_n) \neq 0$ and hence conclude that $ad(Y)$ is nilpotent on $\mathfrak{h}_X$, ie $\mathfrak{h}_X$ is a nilpotent Lie algebra. Deduce from this that if $Y \in \mathfrak{h}_X$, then Y is regular iff $\zeta(Y) = det(ad(Y)|_{\mathfrak{g}/\mathfrak{h}_X}) \neq 0$.

[3] This problem is a prerequisite for attempting the previous problems: let V be a finite dimensional vector space and T a linear operator on V. Let W be a T invariant r-dimensional subspace of V. Choose any basis $\{e_{r+1}+W, ..., e_n+W\}$ for V/W. Show that if $\{e_1, ..., e_r\}$ is any basis for W, then $B = \{e_1, ..., e_n\}$ is a basis for V. Show that $[T]_B$ has the following block structure:

$$[T]_B = \begin{pmatrix} A_{11} & A_{12} \\ 0 & A_{22} \end{pmatrix}$$

and hence deduce that

$$det(T) == det(A_{11}).det(A_{22}) = det(T|_W).det(T|_{V/W})$$

Note that W is the zero element of the vector space V/W. Specialize this result to show that the characteristic polynomial of T can be expressed as

$$f(t) = det(tI - T) = det(tI - T|_W).det(tI - T|_{V/W})$$

1.14 Appendix:Some applications of matrix theory to control theory problems

A.Controllability of the Yang-Mills non-Abelian field equations

The Lagrangian for the non-Abelian gauge fields $A^a_\mu(x), a = 1, 2, ..., N, \mu = 0, 1, 2, 3$ is

$$L = (-1/2)F^a_{\mu\nu}F^{\mu\nu a}$$

where

$$F^a_{\mu\nu} = [D_\mu, D_\nu]^a$$

with D_μ the gauge covariant derivative defined by

$$D_\mu = \partial_\mu + iA^a_\mu\tau_a = \partial_\mu + iA_\mu$$

where $\tau_a, a = 1, 2, ..., N$ are Hermitian generators of the gauge group Lie algebra $\mathfrak{g} = Lie(G)$. The structure constants associated with these generators are denoted by $C(abc)$:

$$[\tau_a, \tau_b] = -i \sum_{c=1}^{N} C(abc)\tau_c$$

Note that

$$F_{\mu\nu} = F^a_{\mu\nu} i\tau_a = [D_\mu, D_\nu] =$$

$$[\partial_\mu + iA_\mu, \partial_\nu + iA_\nu] =$$

$$i(A_{\nu,\mu} - A_{\mu,\nu}) - [A_\mu, A_\nu]$$

$$= (A^a_{\nu,\mu} - A^a_{\mu,\nu})i\tau_a - A^b_\mu A^c_\nu[\tau_b, \tau_c]$$

$$= (A^a_{\nu,\mu} - A^a_{\mu,\nu} + C(abc)A^b_\mu A^c_\nu)i\tau_a$$

Thus,

$$F^a_{\mu\nu} = A^a_{\nu,\mu} - A^a_{\mu,\nu} + C(abc)A^b_\mu A^c_\nu$$

The field equations in the absence of current sources are obtained from the variational principle

$$\delta \int F^a_{\mu\nu} F^{\mu\nu a} d^4x = 0$$

and these give

$$[D_\nu, F^{\mu\nu}] = 0$$

or equivalently,

$$\partial_\nu F^{\mu\nu a} + C(abc)A^b_\nu F^{\mu\nu c} = 0$$

in the presence of interaction with a current source $J^{\mu a}$, with the interaction Lagrangian being

$$L_{int} = J^{\mu a} A^a_\mu$$

the field equations are

$$\partial_\nu F^{\mu\nu a} + C(abc)A^b_\nu F^{\mu\nu c} = J^{\mu a}$$

and these are non-Abelian G generalizations of the Abelian $U(1)$ electromagnetic field equations where G is a Lie subgroup of $U(M)$ with $dimG = N$, where by the dimension of a Lie group, we mean the dimension of its Lie algebra as a vector space. The above field equations be expanded as

$$\partial_\nu(A^{\nu a,\mu} - A^{\mu a,\nu} + C(abc)A^{\mu b}A^{\nu c})$$

$$+C(abc)A^b_\nu(A^{\nu c,\mu} - A^{\mu c,\nu} + C(cde)A^{\mu d}A^{\nu e}) = J^{\mu a}$$

or equivalently after attaching a perturbation parameter δ to keep track of the quadratic and cubic non-linear terms,

$$A^{\nu a,\mu}_{,\nu} - A^{\mu a,\nu}_{,\nu} + \delta C(abc)(A^{\mu b}A^{\nu c}))_{,\nu}$$

$$+\delta C(abc)A^b_\nu(A^{\nu c,\mu}-A^{\mu c,\nu})+\delta^2 C(abc)C(cde)A^b_\nu A^{\mu d}A^{\nu e}=J^{\mu a}---(1)$$

The controllability problem for these field equations is then posed as follows: Let U and V be two disjoint subsets of $\mathbb{R}^3$. Then, given that at time $t=0$, the potentials A^a_μ and their time derivatives $A^a_{\mu,0}$ have prescribed values on U, does there exist a control current field $J^{\mu a}(t,r), 0\le t\le T, r\in\mathbb{R}^3$ such that at time T, these potentials have prescribed values on V ?. More generally, we can ask the question that given two disjoint subsets U,V of $\mathbb{R}^4$, does there exist a control current field $J^{\mu a}(x)$ on $\mathbb{R}^4$ and a solution A^a_μ to the above Yang-Mills field equations corresponding to this current source such that these potentials have prescribed values on both U and V ? We shall attempt to solve this controllability problem approximately by means of perturbation theory. First, we expand the solution in powers of δ:

$$A^a_\mu=A^{a(0)}_\mu+\sum_{k\ge 1}\delta^k.A^{a(k)}_\mu---(2)$$

Since the gauge group G has dimension N equal to the number of possible values of the gauge index a, we can always gauge transform the gauge field so that the gauge conditions

$$\partial_\mu A^{\mu(a)}=0$$

hold good. In that case, (1) reduces to

$$-A^{\mu a,\nu}_{,\nu}+\delta C(abc)A^{\mu b}_{,\nu}A^{\nu c}$$

$$+\delta C(abc)A^b_\nu(A^{\nu c,\mu}-A^{\mu c,\nu})+\delta^2 C(abc)C(cde)A^b_\nu A^{\mu d}A^{\nu e}=J^{\mu a}---(3a)$$

or equivalently with

$$\Box=\partial^\nu\partial_\nu=\partial_0^2-\nabla^2$$

denoting the D'Alembert wave operator,

$$-\Box A^{\mu a}+\delta C(abc)A^{\mu b}_{,\nu}A^{\nu c}$$

$$+\delta C(abc)A^b_\nu(A^{\nu c,\mu}-A^{\mu c,\nu})+\delta^2 C(abc)C(cde)A^b_\nu A^{\mu d}A^{\nu e}=J^{\mu a}---(3b)$$

or on using the antisymmetry of the structure constants,

$$-\Box A^{\mu a}+\delta C(abc)A^{\nu c}(2A^{\mu b}_{,\nu}-A^b_{\nu,\mu})$$

$$+\delta^2 C(abc)C(cde)A^b_\nu A^{\mu d}A^{\nu e}=J^{\mu a}---(3b)$$

Substituting the perturbation expansion (2) into (3b) and equating equal powers of δ gives us (a), for $\delta^0=1$,

$$-\Box A^{\mu a(0)}=J^{\mu a},$$

for $\delta^1=\delta$,

$$-\Box A^{\mu a(1)}+C(abc)A^{\nu c(0)}(2A^{\mu b(0)}_{,\nu}-A^{b(0)}_{\nu,\mu})=0,$$

for δ^2,

$$-\Box A^{\mu a(2)} + C(abc)[A^{\nu c(0)}(2A^{\mu b(1)}_{,\nu} - A^{b(1)}_{\nu,\mu})$$
$$+A^{\nu c(1)}(2A^{\mu b(0)}_{,\nu} - A^{b(0)}_{\nu,\mu})]$$
$$+C(abc)C(cde)A^{b(0)}_{\nu}A^{\mu d(0)}A^{\nu e(0)} = 0$$

Let $G(x-x')$ denote the Green's function for the wave operator $\Box$. Then, we can successively solve the above equations as

$$A^{\mu a(0)}(x) = -\int G(x-x')J^{\mu a}(x')d^4x'$$

$$A^{\mu a(1)}(x) = C(abc)\int G(x-x')A^{\nu c(0)}(x')(2A^{\mu b(0)}_{,\nu} - A^{b(0)}_{\nu,\mu})(x')d^4x',$$

$$A^{\mu a(2)}(x) =$$

$$C(abc)\int G(x-x')[A^{\nu c(0)}(x')(2A^{\mu b(1)}_{,\nu} - A^{b(1)}_{\nu,\mu})(x')$$
$$+A^{\nu c(1)}(x')(2A^{\mu b(0)}_{,\nu} - A^{b(0)}_{\nu,\mu})(x')]d^4x'$$
$$+C(abc)C(cde)\int G(x-x')A^{b(0)}_{\nu}(x')A^{\mu d(0)}(x')A^{\nu e(0)}(x')d^4x'$$

Remark: $G(x)$ satisfies the pde

$$\Box G(x) = \delta^4(x)$$

which on Four dimensional Fourier transforming gives

$$\hat{G}(k) = \frac{1}{k^2}, k^2 = k_\mu k^\mu$$

It is easily deduced then that

$$G(x) = C\delta(x^2), x^2 = x_\mu x^\mu$$

is one such solution. That follows by four dimensional Fourier inversion. It is easily seen from the above formulas that upto $O(\delta^2)$, the solution can be expressed as

$$A(x) = \int G_0(x-x')J(x')d^4x' + \delta\int G_1(x-x', x-y')J(x')\otimes J(y')d^4x'd^4y' +$$

$$+\delta^2\int G_2(x-x', x-y', x-z')(J(x')\otimes J(y')\otimes J(z'))d^4x'd^4y'd^4z' \quad ---(4)$$

where $A(x), J(x)$ are appropriate vector space valued smooth functions on $\mathbb{R}^4$ and G_0, G_1, G_2 are appropriate matrix valued known functions on $\mathbb{R}^4, \mathbb{R}^4\times\mathbb{R}^4$ and $\mathbb{R}^4\times\mathbb{R}^4\times\mathbb{R}^4$. Now the controllability problem is easily stated: Given an $\epsilon > 0$ and a function $A_g(x)$ on $\mathbb{R}^4$, does there exist a source current $J(x)$ such that the output of the above system is $A_d(x)$. Actually, the problem is more intricate if we take initial conditions into account. In that case, we absorb the intial conditions $A_i(0,r)$ into the solution for the zeroth order perturbation. Then,

$$A_0(x) = \int G(x-x')J(x')d^4x' + \int F(t,r|r')A_i(0,r')d^3r'$$

This initial condition then propagates into the higher order perturbations yielding finally upto second order in δ a solution of the form

$$A(x) = \int G_0(x-x')J(x')d^4x' + \delta \int G_1(x-x', x-y')(J(x') \otimes J(y'))d^4x'd^4y' +$$

$$+\delta^2 \int G_2(x-x', x-y', x-z')(J(x') \otimes J(y') \otimes J(z'))d^4x'd^4y'd^4z'$$

$$+ \int F_0(t,r|r')A_i(0,r')d^3r' + \delta \int F_1(t,r|r',r'')A_i(0,r') \otimes A_i(0,r'')d^3r'd^3r''$$

$$+\delta^2 \int F_2(t,r|r',r'')(A_i(0,r') \otimes A_i(0,r'') \otimes A_i(0,r''))d^3r'd^3r''d^3r''$$

and the question of approximate controllability is then the question of whether for a given input field $A_i(r)$ at time 0 and a given output field $A_f(r)$ at time T and an $\epsilon > 0$, does there exist a control input current field $J(x)$ for which the output $A(T,r)$ in the above equation at time T has a mean weighted square distance from $A_f(r)$ defined by

$$\int W(r) \parallel A(T,r) - A_f(r) \parallel^2 d^3r$$

smaller than ϵ ?

Remark: After discretization in the spatial variables, the Yang Mills field equations appears in state variable form as

$$x'(t) = A_0x(t) + \delta.A_1(x(t) \otimes x(t)) + \delta^2.A_2(x(t) \otimes x(t) \otimes x(t)) + u(t)$$

A second order perturbative solution gives

$$x(t) = \int_0^t G_0(t-s)u(s)ds + \delta.\int_0^t \int_0^t G_1(t-s_1, t-s_2)(u(s_1) \otimes u(s_2))ds_1ds_2 +$$

$$\int_{[0,t]^3} G_2(t-s_1, t-s_2, t-s_3)u(s_1) \otimes u(s_2) \otimes u(s_3)ds_1ds_2ds_3$$

where

$$G_0(t) = exp(tA_0)$$

$$G_1(t_1,t_2) =$$

$$G_2(t_1,t_2,t_3) =$$

Using second order perturbation theory, with

$$x(t) = x_0(t) + \delta.x_1(t) + \delta^2.x_2(t) + O(\delta^3)$$

$$x_1'(t) = A_0x_1(t) + A_1(x_0(t) \otimes x_0(t))$$

so

$$x_1(t) = \int_{0<s_1,s_2<s<t} G_0(t-s)A_1(G_0(s-s_1)\otimes G_0(s-s_2))(u(s_1)\otimes u(s_2))ds_1ds_2ds$$

and hence

$$G_1(t-s_1,t-s_2) = \int_{max(s_1,s_2)<s<t} G_0(t-s)A_1(G_0(s-s_1)\otimes G_0(s-s_2))ds$$

$$x_2'(t) = A_0x_2(t) + A_1(x_0(t)\otimes x_1(t) + x_1(t)\otimes x_0(t)) + A_2(x_0(t)\otimes x_0(t)\otimes x_0(t))$$

so that

$$x_2(t) = \int_0^t G_0(t-s)(A_1(x_0(s)\otimes x_1(s)+x_1(s)\otimes x_0(s))+A_2(x_0(s)\otimes x_0(s)\otimes x_0(s))ds$$

Remark: If the input $u(t)$ is replaced by $Bu(t)$ where B is a rectangular matrix, then

$$x_0(t) = \int_0^t G_0(t-s)Bu(s)ds,$$

$$x_1(t) = \int_0^t\int_0^t G_1(t-s_1,t-s_2)(B\otimes B)(u(s_1)\otimes u(s_2))ds_1ds_2$$

It is clear from these expressions that for controllability upto $O(\delta)$, we must have that the matrix

$$C = [A^kB,, A_0^rA_1(A_0^s\otimes A_0^m)(B\otimes B), k,r,s,m \geq 0]$$

must have full row rank. Likewise a necessary condition for controllability upto $O(\delta^2)$ can be formulated.

Remark: It should be noted that in view of the Cayley-Hamilton equation satisfied by A_0, it is enough to restrict all the indices k, r, s, m in the above controllability matrix C to assume values in the range $0, 1, ..., n-1$ where $A_0 \in \mathbb{R}^{n\times n}$.

Appendix

Let $A_\mu \to A'_\mu$ be the transformation of the potential under a gauge group element $g(x) \in G$ dependent upon space-time in a differentiable way. Then, if $\psi(x)$ denotes the matter field which appears in the matter component of the Lagrangian either as a function of the ψ, ψ^* and $D_\mu\psi, (D_\mu\psi)^*$, then it is clear that since under this local gauge transformation $\psi(x) \to g(x)\psi(x) = \psi'(x)$, in order to maintain invariance of the matter component of the Lagrangian under local gauge transformations, it must be of the form $L_M(\psi(x), D_\mu\psi(x))$ where L_M is G-invariant, ie,

$$L_M(g.\psi, g.\xi_\mu) = L_M(\psi, \xi_\mu), g \in G$$

and hence for $L_M(\psi(x), D_\mu\psi(x))$ to be locally G-invariant, the potential A_μ must transform to A'_μ so that if $D'_\mu = \partial_\mu + A'_\mu$, then

$$D'_\mu g(x)\psi(x) = D'_\mu\psi'(x) = g(x).D_\mu\psi(x)$$

This happens only provided that

$$D'_\mu = g(x).D_\mu.g(x)^{-1}$$

or equivalently,

$$A'_\mu(x) = g(x)A_\mu(x).g(x)^{-1} + g(x)(\partial_\mu g(x)^{-1})$$

$$= g(x)A_\mu(x).g(x)^{-1} - (\partial_\mu g(x))g(x)^{-1}$$

In particular, for an infinitesimal local gauge transformation,

$$g(x) = 1 + \epsilon^a(x)\tau_a = 1 + \epsilon(x)$$

we get upto $O(\epsilon)$, the infinitesimal gauge transformation,

$$A'_\mu(x) - A_\mu(x) = \delta A_\mu(x) =$$

$$[\epsilon(x), A_\mu(x)] - \epsilon_{,\mu}(x)$$

or in component form,

$$\delta A^a_\mu(x) = -\epsilon^a_{,\mu}(x) + C(abc)\epsilon^b(x)A^c_\mu(x)$$

One candidate for the total action functional for the non-Abelian gauge and matter fields is given by

$$S[\psi, A_\mu] = (-1/4)\int F^a_{\mu\nu}F^{\mu\nu a}d^4x + c_1\int[\psi^*\gamma^0(i\gamma^\mu D_\mu - m)\psi]d^4x$$

$$+c_2\int(D^\mu\chi)^*(D_\mu\chi)d^4x$$

where χ is a complex Klein-Gordon scalar field while ψ is a non-Abelian Dirac matter field. This action is invariant under local gauge transformations. Note that the gauge group transformations commute with the Dirac Gamma matrices. The local gauge group symmetry can be broken by introducing control terms like $\int A^a_\mu J^{\mu a}d^4x$ and $\int B^{\mu a}\psi^*\gamma^0\gamma^\mu\psi d^4x$ with the control current $J^{\mu a}$ and the control gauge field potentials $B^{\mu a}$ being classical, ie, c-number fields. It should be noted here that $\psi^*\gamma^0\gamma^\mu\psi$ is the non-Abelian Dirac current density. Likewise, the Klein-Gordon current is

$$(-i/2)[\chi^* D_\mu\chi - \chi(D_\mu\chi)^*] = Im(\chi^* D_\mu\chi)$$

and we can also introduce an interaction term between this current and a control c-number four potential field C^μ as

$$\int C^\mu Im(\chi^* D_\mu \chi) d^4x$$

which also breaks the local gauge symmetry of the action.

Remark: This controllability problem is a generalization to the four dimensional continuum of the following problem in ordinary differential equations: Consider the vector valued ode

$$\frac{dx(t)}{dt} = A_0 x(t) + A_1(x(t) \otimes x(t)) + .. + A_m(x(t)^{\otimes m}) + Bu(t), t \geq 0$$

where $x(t) \in \mathbb{R}^n$. Then under what conditions on the matrices $A_0, .., A_m, B$ is this system controllable ?, ie, when does there exist a control input $u(t), 0 \leq t \leq T$ that takes any given initial state $x(0)$ to a final state $x(T)$? If $A_k = 0, k = 1, 2, ..., m$, then Kalman proved that the system is controllable iff the matrix

$$[B, AB, A^2B, ..., A^{n-1}B]$$

has full row rank.

Symmetry breaking in non-Abelian gauge field theory. Let the gauge group G be spontaneously broken into the subgroup H. Thus, if $\psi(x)$ transforms according to G, and if we write $g(x) = \gamma(x)h(x)$ where $h(x) \in H$ and $\gamma(x)$ is a representative element for the coset space G/H, then we can express the local gauge transformation of $\psi(x)$ as follows: First write

$$\psi(x) = \gamma(x)\tilde{\psi}(x)$$

where $\tilde{\psi}(x)$ transforms according to H. Then if $g \in G$, the transformation law of ψ under g can be expressed as

$$\psi(x) \to \psi'(x) = g\psi(x) = g\gamma(x)\tilde{\psi}(x) =$$

$$\gamma(g, x)h(g, x)\tilde{\psi}(x)$$

where

$$h(g, x) \in H,$$

and $\gamma(g, x)$ is the representative element in the coset space G/H corresponding to the element $g\gamma(x) \in G$. Thus, if we write

$$\psi'(x) = \gamma'(g, x)\tilde{\psi}'(x)$$

where $\gamma'(g, x)$ is a representative element in the coset space G/H and $\tilde{\psi}'(x)$ transforms according to H, then we have

$$\gamma'(g, x)\tilde{\psi}'(x) = \gamma(g, x)h(g, x)\tilde{\psi}(x)$$

Note that

$$g\gamma(x) = \gamma(g,x)h(g,x)$$

Remark: We can look upon $\gamma(x)$ as that component of the wave function $\psi(x)$ that corresponds to the broken symmetry degrees of freedom and $\tilde{\psi}(x)$ as that component of $\psi(x)$ that transforms according to the unbroken subgroup H. Note that the broken degrees of freedom are represented by the coset space G/H. Consider now the gauge transformation of the gauge field $A_\mu(x)$. $D_\mu = \partial_\mu + iA_\mu$ transforms under local gauge transformations according to the adjoint representation of the gauge group. Now define $\tilde{A}_\mu(x)$ by the equation

$$\tilde{A}_\mu(x) = \gamma(x)^{-1}A_\mu(x)\gamma(x)$$

or equivalently,

$$A_\mu(x) = \gamma(x)\tilde{A}_\mu(x)\gamma(x)^{-1}$$

This transformation may be viewed as the removal of the broken symmetry degrees of freedom from the gauge potentials A_μ. We therefore expect $\tilde{A}_\mu(x)$ to transform under local gauge transformations $g(x) \in G$ in accordance with an unbroken subgroup H element derived from $g(x)$. more precisely, for a given gauge group transformation $g(x) \in G$, we expect that $\tilde{D}_\mu = \partial_\mu + i\tilde{A}_\mu(x)$, after perhaps the addition of some space-time element in the Lie algebra of H, to transform according to $Ad(h(g,x))$. To make these matters precise, we compute:

1.15 Controllability of supersymmetric field theoretic problems

The Lagrangian of the super-Yang-Mills field is

$$L = F^a_{\mu\nu}F^{\mu\nu a} + c.\bar{\chi}^a\gamma^\mu D_\mu\chi^a - -- (A.1)$$

where $F^a_{\mu\nu}$ is the antisymmetric field tensor corresponding to the Yang-Mills gauge potentials A^a_μ and χ^a is the gaugino field. D_μ acts on the gaugino field in the adjoint representation. In other words, by $D_\mu\chi^a$ we actually mean $[D_\mu,\chi]^a$ where $\chi = \chi^a\tau_a$ and $D_\mu = \partial_\mu + iA^a_\mu\tau_a$. Thus,

$$[D_\mu,\chi]^a = \chi^a_{,\mu} + C(abc)A^b_\mu\chi^c$$

It is well known that the action corresponding to the Lagrangian (1) is invariant under the infinitesimal local supersymmetric transformations

$$\delta A^a_\mu(x) = \bar{\epsilon}(x)\gamma_\mu\chi^a(x),$$

$$\delta\chi^a(x) = \gamma_{\mu\nu}\epsilon(x)F^{\mu\nu a}(x)$$

where $\epsilon(x)$ is an infinitesimal Majorana parameter field and

$$\bar{\epsilon}(x) = \epsilon(x)^T \gamma^0$$

provided that the dimension of space-time is 10. This critical dimension is needed for the term involving a product of three gaugino Majorana spinor fields χ^a in

$$\begin{aligned}\delta_A(\bar{\chi}^a \gamma^\mu D_\mu \chi^a) &= \bar{\chi}^a \gamma^\mu (\delta D_\mu) \chi^a \\ &= \bar{\chi}^a \gamma^\mu (C(abc)(\delta A^b_\mu)\chi^c \\ &= C(abc)\bar{\chi}^a \gamma^\mu \chi^c . \bar{\epsilon}(x) \gamma_\mu \chi^b(x), \\ &= C(abc)\bar{\chi}^a \gamma^\mu \chi^c \bar{\chi}^b \gamma_\mu \epsilon\end{aligned}$$

to vanish for any $\epsilon(x)$. This is equivalent to require the vanishing of

$$C(abc)\bar{\chi}^a \gamma^\mu \chi^c \bar{\chi}^b \gamma_\mu$$

Then, to the locally supersymmetric Lagrangian (1), we can add supersymmetry breaking terms by coupling the gauge field A^a_μ to classical current source $J^{\mu a}$ and its superpartner, the gaugino field χ^a to a classical vector potential C^μ so that the perturbing Lagrangian is

$$\Delta L = J^{\mu a} A^a_\mu + C^\mu \bar{\chi}^a \gamma_\mu \chi^a$$

and then address the controllability question.

1.16 Controllability of Yang-Mills gauge fields in the quantum context using Feynman's path integral approach to quantum field theory

Consider the action functional

$$I_T[A|J] = \int_{0 \le x_0 \le T, (x^1, x^2, x^3) \in \mathbb{R}^3} [(-1/4) F^a_{\mu\nu} F^{\mu\nu a} + A^a_\mu J^{\mu a}] d^4x$$

where $J^{\mu a}(x)$ is a classical current field. Let $\psi_\alpha(A), \alpha \in I$ be a family of wave functionals of the gauge field A and let S_g be a given scattering matrix in gauge field space. We compute the scattering matrix elements w.r.t these states:

$$< \psi_\alpha | S_g | \psi_\beta >= \int \psi_\alpha(A)^* S_g \psi_\beta(A) > dA, \alpha, \beta \in I$$

and ask the question, when does there exist a control current field J so that the actual scattering matrix elements over a given time duration $[0, T]$ defined by

$$S_{\alpha\beta}(J) = \int \psi_\alpha(A)^* S(J) \psi_\beta(A) dA$$

where

$$S(J)\psi(A_f) = \int K(A_f, A_i|J)\psi(A_i)dA_i$$

with

$$K(A_f, A_i|J) = \int_{A(0,.)=A_i, A(T,.)=A_f} exp(iI_T[A|J])\Pi_{0\leq x^0\leq T, \mathbf{x}\in\mathbb{R}^3} dA(x^0, \mathbf{x})$$

being the Feynman path integral evolution kernel are close to the given ones ?

1.17 Large deviations and control theory

Let $p(\partial) = L$ be a linear partial differential operator in n variables $x = (x_1, ..., x_n)$ such that L transforms a vector valued signal $[f_1(x), ..., f_n(x]^T$ into another vector valued signal $[g_1(x), ..., g_m(x)]^T$. Thus, L can be looked upon as an $m \times n$ matrix partial differential operator:

$$L = ((p_{ij}(\partial)))_{1\leq i\leq m, 1\leq j\leq n}$$

We consider the special case when $m = n$ and then consider the stochastic pde

$$Lf(x) = s(x) + \sqrt{\epsilon}w(x), x \in \mathbb{R}^n$$

where $s(x)$ is an input signal field and $w(x)$ is a zero mean Gaussian noise field with covariance

$$\mathbb{E}(w(x)w(y)^T) = K_w(x, y)$$

In the absence of noise, assuming L to be invertible, we write the general solution as $f_0(x) = L^{-1}s(x) + h(x)$ where $h(x)$ is any solution to the homogeneous pde

$$Lh(x) = 0$$

Now we ask the following question: When weak noise is present, then what is the approximate probability for the solution

$$f(x) = L^{-1}(s(x) + \sqrt{\epsilon}w(x)) + h(x) = f_0(x) + \sqrt{\epsilon}L^{-1}(w)(x)$$

to deviate by an amount greater than δ from the noiseless solution $f_0(x)$ over a domain $D \subset \mathbb{R}^n$ in the L^2 sense ? Using the theory of large deviations for Gaussian random variables, this probability is approximately given by

$$P(\int_D \| f(x) - f_0(x) \|^2 > dx\delta^2) \approx$$

$$exp(-(1/2(\epsilon))inf_{g:\int_D\|g(x)\|^2>\delta^2} \int_{D\times D} g(x)^T Q(x, y)g(y)dxdy)$$

where

$$Q = R^{-1}, R(x,y) = \mathbb{E}((L^{-1}w(x)).(L^{-1}w)(y)^T)$$

$$= \int_{D\times D} L^{-1}(x,u)K_w(u,v)L^{-1}(y,v)dudv$$

This approximate probability evaluates to

$$exp(-\delta^2/2\epsilon\lambda_{max}(R))$$

where $\lambda_{max}(R)$ is the maximum eigenvalue of the kernel R on $D \times D$ and Q is the inverse kernel of R on $D \times D$:

$$\int_D R(x,u)Q(u,y)du = \delta^n(x-y), x,y \in D$$

Now we wish to reduce this deviation probability by adding non-random control terms which would reduce the effects of noise. Specifically, we modify the system to

$$Lf(x) = s(x) + \sqrt{\epsilon}w(x) + u(x,\theta)$$

where $u(x,\theta)$ is an input depending upon a control parameter vector θ. We write

$$L^{-1}u(x,\theta) = v(x,\theta)$$

and then the controlled system output can be expressed as

$$f(x) = f_0(x) + v(x,\theta) + \sqrt{\epsilon}L^{-1}w(x)$$

and the corresponding deviation probability then gets modified to on defining the output noise field

$$d(x) = L^{-1}w(x)$$

to

$$P(\int_D \| f(x) - f_0(x) \|^2 > dx\delta^2)$$

$$= P(\int_D \| v(x,\theta) + \sqrt{\epsilon}d(x) \|^2 \; dx > \geq \delta^2) =$$

$$exp(-(2\epsilon)^{-1}inf_{g:\int_D \|v(x,\theta)+g(x)\|^2 > dx\delta^2} \int_{D\times D} g(x)^T Q(x,y)g(y)dxdy)$$

Finally, the parameter vector θ must be chosen so that this deviation probability is as small as possible, or equivalently, such that

$$inf_{g:\int_D \|v(x,\theta)+g(x)\|^2 dx > \delta^2} \int_{D\times D} g(x)^T Q(x,y)g(y)dxdy)$$

is as large as possible.

Some additional remarks on controllability: Suppose we have a nonlinear vector sde of the form

$$dx(t) = f(t,x(t)|\theta)dt + \sqrt{\epsilon}g(t,x(t),\theta)dB(t)$$

Then, we wish to design the control parameters θ such that the probability of the trajectory falling in a set $D \subset \mathbb{C}[0,T]^n$ over the time duration $[0,T]$ is minimized. Then, the large deviation solution to this problem will be to maximize

$$E(\theta) = \inf_{\{x \in D, \xi \in C^1[0,T]: dx(t)/dt = f(t,x(t),\theta) + g(t,x(t),\theta)\xi(t)\}} \int_0^T \| \xi(t) \|^2 dt$$

1.18 Approximate controllability of the Maxwell equations

The wave equations for the vector and scalar potentials are

$$\Box A^\mu = J^\mu + \sqrt{\epsilon} w^\mu$$

where

$$J^\mu_{,\mu} = 0, w^\mu_{,\mu} = 0$$

and $w^r(x)$ are jointly Gaussian noise fields with zero mean. Thus,

$$w^0_{,0} = -w^r_{,r}, w^0 = -\int_0^{x^0} w^r_{,r} dx^0$$

We now incorporate control terms in the current, ie, replace J^μ by $J^\mu(x) + K^\mu(x|\theta)$ where θ are control parameters and K^μ satisfies

$$K^\mu_{,\mu}(x|\theta) = 0 \forall \theta \in \Theta$$

The solution is

$$A^\mu(x) = \int G(x-x')(J^\mu(x') + K^\mu(x'|\theta) + \sqrt{\epsilon} w^\mu(x'|\theta)) d^4x'$$

and the problem is to use large deviation theory for Gaussian processes to mininimize the probability

$$P(\int_D \| A^\mu(x) - A^\mu_d(x) \|^2 W(x) d^4x > \epsilon)$$

by instead minimizing the inifimum of the rate function of A^μ over the given set indicated in the probability w.r.t the control parameters θ.

1.19 Controllability problems in quantum scattering theory

Let H_0 denote the free projectile Hamiltonian and $H_1(\theta)$ the Hamiltonian when the projectile interacts with the scattering centre. θ is a control parameter vector for the scattering potential. The wave operators are

$$\Omega_+(\theta) = lim_{t\to\infty}exp(-iH_1(\theta)).exp(itH_0),$$

$$\Omega_-(\theta) = lim_{t\to-\infty}exp(-itH_1(\theta)).exp(itH_0)$$

These wave operators can be computed using the Lippmann-Schwinger equations. The scattering matrix is then

$$S(\theta) = \Omega_+(\theta)^*\Omega_-(\theta)$$

The controllability problem is then to determine whether for each given scattering operator S_g (ie, a unitary operator) in a given family whether there exists a θ for which $S(\theta)$ has a distance smaller than ϵ w.r.t the spectral norm from S_d.

1.20 Kalman's notion of controllability and its extension to pde's

Consider the state equations

$$X'(t) = AX(t) + BU(t)$$

This can also be expressed as

$$[Id/dt - A, B]\begin{pmatrix} X(t) \\ U(t) \end{pmatrix} = 0$$

The controllability problem then involves determining whether a solution $(X(t), U(t))$ to this equation exists over the time interval $[t_1, t_2]$ such that the value of $X(t)$ at t_1 and t_2 are given. Generalizing this to pde's, we ask the question, given a matrix partial differential operator $p(\partial)$, there exists a solution f to it, ie, $p(\partial)f(x) = 0$ such that $\Psi(f(x))$ has specified values on two disjoint Borel sets U and V. Note that in the above special case considered by Kalman, the two disjoint open sets are $\{t_1\}$ and $\{t_2\}$.

1.21 Controllability in the context of representations of Lie groups

Let π be a representation of a Lie group G that acts on a manifold M. Let an image field $f_1(x)$ be given on M. After transforming it by a G-action and adding noise to it, the image field becomes

$$f_2(x) = f_1(g^{-1}.x) + w(x), x \in M$$

We can regard f_1 as the input image field and f_2 as the output image field with the system being defined by $g \in G$.

We can write $g(t) = exp(tX)$ for a one parameter sub-group $t \to g(t)$ of G where X is an element of the Lie algebra $\mathfrak{g}$ of G. Then the initial image field $f(x)$ after time t transforms to

$$f_2(t, x) = f_1(exp(-tX).x), t \geq 0, x \in M$$

Its rate of change at time t is given by

$$\partial f_2(t, x)/\partial t = -\xi_X(x).f_2(t, x)$$

where ξ_X is the vector field induced on M by the infinitesimal action of the one parameter group $g(t)$ on M, ie,

$$\xi_X(x) = \frac{d}{dt} exp(tX).x|_{t=0}$$

Formally, we can express the solution to the above pde as

$$f_2(t, x) = exp(-t\xi_X(x)).f_1(x)$$

and now we can pose the controllability question: Given a family of smooth functions $\mathcal{F}$ on M, and two elements f_a, f_b in M does there exist an element $X \in \mathfrak{g}$ such that

$$f_b(x) = exp(-T.\xi_X(x))f_a(x)$$

for some fixed $T > 0$. Note that this is equivalent to the existence of a one parameter group $g(t)$ such that

$$f_b(x) = f_a(g(T)^{-1}x)$$

Now the representation π of G induces a representation $d\pi$ of $\mathfrak{g}$ such that

$$\pi(exp(tX)) = exp(t\pi(X)), t \geq 0$$

If G acts transitively on M, we can define a Fourier transform of $f(x)$ at π formally by choosing a point $x_0 \in M$ and defining

$$\hat{f}(\pi) = \int_G f(gx_0)\pi(g)dg$$

After time t, $f(x)$ evolves to $f(t,x) = f(exp(-tX)x)$ and its Fourier transform at π is then given by

$$\hat{f}(t,\pi) = \int_G f(t, gx_0)\pi(g)dg = \int_G f(exp(-tX)gx_0)\pi(g)dg$$

$$= \int_G f(gx_0)\pi(exp(tX)g)dg = \pi(exp(tX))\hat{f}(\pi)$$

which is equivalent to saying that

$$\partial\hat{f}(t,\pi)/\partial t = d\pi(X).\hat{f}(t,\pi)$$

the right side being intepreted in terms of ordinary matrix multiplication. Thus, we have

$$\hat{f}(t,\pi) = exp(t.d\pi(X))\hat{f}(\pi)$$

and hence we can pose the controllability problem of when this operation will carry the Fourier transform of an initial signal field evaluated at π to the Fourier transform of another signal field at π when both the signal fields are taken from a given family ?

1.22 Irreducible representations and maximal ideals

Let A be an algebra with a unit 1 and π a representation of this algebra in a vector space V. Assume that π has a cyclic vector v, ie,

$$\pi(A)v = V$$

Define

$$I_0 = \{x \in A : \pi(x)v = 0\}$$

Clearly, I_0 is a left ideal of A and A/I_0 is isomorphic to V as a vector space via the mapping $x + I_0 \to \pi(x)v, x \in A$. Now, let W be a π-invariant subspace of V and define

$$I_W = \{x \in A : \pi(x)v \in W\}$$

Clearly since W is π invariant, it follows that I_W is a left ideal in A containing I_0. We can define a representation $\bar{\pi}$ of A in A/I_0 by $\bar{\pi}(x)(y+I_0) = xy+I_0, x, y \in A$. Then, it is clear that $\bar{\pi}$ is isomorphic to π and that I_W/I_0 is a $\bar{\pi}$-invariant subspace of A/I_0. This argument shows that there is a one-one correspondence, ie, bijection between all π-invariant subspaces of V and all $\bar{\pi}$ invariant subspaces of A/I_0. Equivalently, there is a one-one correspondence between all π-invariant subspaces of V and all ideals of A containing I_0. In particular, if π is irreducible, then I_0 is a maximal ideal of A and conversely if I_0 is any maximal ideal of A, then the representation $\bar{\pi}$ of A in A/I_0 is irreducible with $1+I_0$ as a cyclic vector. In other words, there is a one-one correspondence between equivalence classes of irreducible representations of A and maximal ideals in A. This fact was used by Harish-Chandra with great power in constructing all the finite dimensional irreducible representations of a semisimple Lie algebra using dominant integral weights.

1.23 Controllability of the Maxwell-Dirac equations using external classical current and field sources

Suppose that we are given a second quantized Dirac electron-positron field and and a second quantized Maxwell photon field inside a cavity having perfectly conducting boundary. We wish to control these fields so that the far field radiation pattern has a given set of quantum moments in a given state of the cavity field, say the tensor product of a Bosonic and Fermionic coherent state. Let A_μ^c and J_μ^c denote the external classical field and current sources into the cavity. The perturbed Maxwell-Dirac equations are then given by

$$\square A_\mu = -e\psi^* \alpha^\mu \psi + J_\mu^c,$$

$$\gamma^\mu (i\partial_\mu - m)\psi = -e\gamma^\mu (A_\mu + A_\mu^c)\psi$$

Let $D(x - x')$ and $S(x - x')$ denote respectively the photon and electron propagator kernels:

$$D = \square^{-1}, S = (i\gamma^\mu \partial_\mu - m + i0)^{-1}$$

The approximate solutions to these equations is

$$A_\mu(x) = A_\mu^{(0)}(x) + \int G(x - x')(-e\psi_0(x')^* \alpha_\mu \psi_0(x') + J_\mu^c(x'))d^4x'$$

$$= A_\mu^{(0)}(x) + \delta A_\mu(x)$$

$$\psi(x) = \psi_0(x) + \int S(x - x')(-e\gamma^\nu (A_\nu^{(0)}(x') + A_\nu^c(x'))d^4x'$$

$$= \psi_0(x) + \delta\psi(x),$$

where ψ_0 free second quantized Dirac field that satisfies

$$[i\gamma^\mu \partial_\mu - m]\psi_0 = 0$$

and $A_\mu^{(0)}$ is the free second quantized photon field that satisfies

$$\square A_\mu^{(0)} = 0$$

ψ_0 is expressed as a linear superposition of plane waves with dispersion relation $p^0 = \sqrt{m^2 + p_1^2 + p_2^2 + p_3^2}$ with coefficients being the electron annihilation and positron creation operators in momentum space while $A_\mu^{(0)}$ is expressed as a linear superposition of plane waves with dispersion relation $k^0 = \sqrt{k_1^2 + k_2^2 + k_3^2}$ with coefficients being photon annihilation and creation operators in momentum space. Using these expressions, if $|\Phi>$ is any state of the electrons, positrons and photons within the cavity, then we can calculate in principle all the moments of the radiation and Dirac field in this state. electromagnetic radiation from the cavity comes from the current of electron and positrons within the cavity as well as from the surface current density induced by the the quantum magnetic field on the cavity boundary. Applying the retarded potential formula to these two currents, it follows that the total quantum electromagnetic field radiated from the cavity will have the form

$$A_{rad,\mu}(x) = \int_{cavity} G_1(x,x')\psi(x')^*\alpha_\mu\psi(x')d^4x'$$

$$+ \int_{cavity} G_{2\mu}^{\nu}(x,x')A_\nu(x')d^4x'$$

In this expansion, we retain terms only upto linear orders in $\delta\psi$ and δA_μ. It follows then that we can express the radiated electromagnetic field as

$$A_{rad,\mu}(x) = F_{1\mu}(x) + \int F_{2,\mu}(x,x')\delta\psi(x')d^4x'$$

$$+ \int F_{3\mu}^{\nu}(x,x')\delta A_\nu(x')d^4x'$$

where $F_{2mu}(x,x')$ is a linear functional of ψ_0 and hence of the electron-positron creation and annihilation operators while $F_{3\mu}^{\nu}$ are c-number functions. $F_{1\mu}$ is the retarded Maxwell potential produced by the free Dirac quantum current density $-e\psi_0^*\alpha_\mu\psi_0$ and is therefore a quadratic functional of the electron and positron creation and annihilation operator fields. It should be noted that the classical current and potential sources are contained in the terms $\delta\psi$ and δA_μ.

The controllability issue can then be posed as follows: For a given $\epsilon, \delta > 0$ and a given electromagnetic field $A_{g,\mu}(x')$ in a region V of space-time, do there exist classical control fields J_μ^c and A_μ^c so that the quantum average of $A_{rad,\mu}(x)$ in the given coherent state of the electrons, positrons and photons has a distance smaller than ϵ from the given electromagnetic field in the sense of a weighted integral of the error square over V and simultaneously, this field has a fluctuation mean square value smaller than δ^2 over this region in the coherent state ? By fluctuation mean square value, we mean the quantity

$$\int_{V\times V} <\Phi|A_{rad,\mu}(x)A_{rad,\nu}(x')|\Phi> W^{\mu\nu}(x,x')d^4xd^4x'$$

$$-\int_{V\times V} <\Phi|A_{rad,\mu}(x)|\Phi><\Phi|A_{rad,\nu}(x')|\Phi> W^{\mu\nu}(x,x')d^4xd^4x'$$

We require that this must be smaller than δ^2 and likewise, we require that

$$\int_{V\times V}(<\Phi|A_{rad,\mu}(x)|\phi> -A_{g,\mu}(x))(<\Phi|A_{rad,\nu}(x')\phi> -A_{g,\nu}(x'))W^{\mu\nu}(x,x')d^4xd^4x'$$

must be smaller than ϵ^2.

1.24 Controllability of the EEG signals on the brain surface modeled as a spherical surface by influencing the infinitesimal dipoles in the cells of the brain cortex to vary in accord to sensory perturbations

If $U(r)$ is the potential generated on the brain surface by infinitesimal dipoles $p_1,...,p_N$ present at the locations $r_1,...,r_N$ in the cortex, then U satisfies Poisson's equation

$$\nabla^2U(r) = -\rho(r)/\epsilon, \rho(r) = \sum_k p_k.\nabla\delta^3(r-r_k)$$

More generally, if we assume the presence of stochastic perturbation terms in the charge distribution, we obtain the following stochastic pde

$$\nabla^2U(r) = -\sum_k p_k.\nabla\delta^3(r-r_k) + w(r)$$

Assume that there are no charges present on the brain surface, ie, if $\hat{n}$ is the unit normal at any point on the brain surface, then

$$\partial U(r)/\partial\hat{n} = 0$$

Then if $G(r,r')$ is the Green's function for the Neumann boundary value problem, we have

$$\nabla^2G(r,r') = \delta^3(r-r'), \partial G(r,r')/\partial\hat{n} = 0$$

and we get as solution

$$U(r) = \int G(r,r')(-\sum_k p_k.\nabla\delta^3(r'-r_k) + w(r'))d^3r'$$

Assume that $w(r)$ is weak zero mean Gaussian noise with known autocorrelation

$$R_w(r,r') = \mathbb{E}(w(r)w(r'))$$

Then, the controllability problem is to add additional charge sources to the brain cortex, defined by a control charge density $\rho_c(r)$, so that the controlled potential on the brain surface

$$U_c(r) = \int G(r,r')(-\sum_k p_k.\nabla\delta^3(r'-r_k) + w(r') + \rho_c(r'))d^3r'$$

has a minimal probability of determining an electric field $E_c(r) = \nabla U_c(r)$ on the brain surface S that deviates from a given surface electric field $E_g(r)$ on S by an amount $> \epsilon$. This is the classical control problem based on large deviation theory.

1.25 Control and relativity

Application of the representation theory of $SL(2,\mathbb{C})$ as an alternative way of characterizing Lorentz transformations to control problems. We use the irreducible representations of $SL(2,\mathbb{C})$ to estimate a Lorentz group transformation element on a time varying three dimensional image field and using this estimate, to design an error feedback controller in the group domain so that the transformed image field is as close as possible in the sense of some distance measure to a given 3-D time varying image field. This idea can be compared to the extended Kalman filter based state observer to design a controller based on output error feedback so that the state tracks a given trajectory.

Given a Lie group or more generally an algebraic group over a field, the question is how to construct an appropriate basis for an irreducible representation of the group or an irreducible representation of a module. The standard Verma-Module or Borel-Weil method involves starting with a formal vector, called the highest weight vector and operating on it freely by the negative root vectors of the semisimple Lie algebra and then extracting out a Maximal ideal from this free module and using the fact that the quotient of the universal enveloping algebra by a maximal ideal is an irreducible module for the universal enveloping algebra of the Lie algebra of the group. Standard monomials based on Schubert varieties of the Grassmannian provide nice bases for irreducible modules of algebraic groups like $SL(n,K), SO(n,K)$ where K is any algebraic field. Such fields naturally appear in the construction of classical codes and we can look upon the elements of these algebraic groups as linear transformations on the space of code vectors and formulate code pattern recognition problems for the same via the irreducible representations of these groups.

Acknowledgements: I am grateful to Professor Shiva-Shankar for suggesting this problem to me and for providing me with his lecture notes on controllability, partial differential equations and the vector potential delivered at the Steklov Institute, Moscow.

References:

[1] Shiva Shankar, "Six lectures at the Steklov Institute, Moscow on Controllability and the vector potential.

[2] Amir Dembo and Ofer Zeitouni, "Large deviations, Techniques and Applications", Springer.

[3] V.S.Varadarajan, "Lie groups, Lie algebras and their Representations, Springer, 1984.

[4] C.S.Seshadri, "An Introduction to Standard Monomials", Hindustan Book Agency.

Chapter 2

Probability Theory

[0] Some philosophical remarks on probability theory: Why probabilistic models are required to simplify calculations involving very complex deterministic dynamical systems ? The Buffon Needle problem and its application to the Monte-Carlo calculation of π.

[1] A.N.Kolmogorov's axiomatic foundations of probability theory: The sample space, σ-algebra of events and probability measure; the notion of a classical probability space $(\Omega, \mathcal{F}, P)$. Importance of the countable additivity postulate for the probability measure.

[2] Properties of the probability measure.

[3] The notion of independence of events. The Borel-Cantelli lemmas.

[4a] The general definition of a random variable on a probability space. Joint probability distributions and their properties.

[4b] Lebesgue integration in a probability space: The notion of expectation of a random variable.

[4c] Cornerstone theorems of Lebesgue integration theory: Monotone convergence theorem, Fatou's Lemma, dominated convergence theorem.

[5] Statement of the Caratheodory extension theorem: Extension of a probability measure as a countably additive measure on an algebra of events to the σ-algebra generated by the algebra.

[6] The product of probability spaces: Notion of the product measure and its application to the construction of independent experiments, Fubini's theorem on integration w.r.t a product probability measure.

[7] Examples of probability spaces from die throwing to coin tossing.

[8] Absolute continuity of two measures and the Radon-Nikodym theorem.

[9] Application of the Radon-Nikodym theorem to the construction of the conditional expectation of a random variable given a sub σ-algebra.

[10] Another derivation of the conditional expectation using orthogonal projection operators in a Hilbert space.

[11] Properties of the conditional expectation.

[12] Application of the conditional expectation to the construction of the minimum mean square nonlinear estimate of a random variable given a family of random variables.

[13] Application of the Radon-Nikodym theorem to the construction of probability density of a finite set of random variables.

[14] Describing discrete probability distributions using the Dirac δ-distribution.

[15] Estimation of parameters in linear models using linear minimum mean square methods.

[16] Joint characteristic function of a finite set of random variables.

[17] Positive definite properties of the characteristic function and Bochner's theorem.

[18] Jensen's inequality for convex functions of random variables.

[19] Chebyshev's inequality, Markov's inequality.

[20] [a] Various notions of convergence of an infinite sequence of random variables. Convergence almost surely, convergence in probability, convergence in the mean square sense, convergence in L^p-norm, convergence in distribution. The relationship between these modes of convergence.

[b] The weak and strong laws of large numbers for sequences of independent random variables.

[c] The Gaussian distribution and the central limit theorem: Proof based on the use of the characteristic function.

[21] Definition of a stochastic process in discrete time and continuous time.Kolmogorov's existence theorem for stochastic processes in discrete time and in continuous time.

[22] The AR,MA and ARMA time series models.

[23] Transmission of stochastic processes through linear and nonlinear filters. Derivation of differential and difference equations satisfied by the output moments and the input output cross moments in terms of the input moments.

[24] Stationary and Wide sense stationary processes.

[25] Von-Neumann's L^2-ergodic theorem, Birkhoff's individual ergodic theorem and ergodicity of a measure preserving transformation with applications to stochastic processes.

[26] Autocorrelation, spectrum, higher order spectra and the causal and non-Causal Wiener filters.

[27] Nonlinear filtering and the Kalman and extended Kalman filters for real time filtering.

[28] Simulation of random variables on a computer by transformation of a uniformly distributed random variable.

[29] The Brownian motion, Poisson process and some of their properties.

[30] Stochastic integration w.r.t Brownian motion and stochastic differential equations driven by the Brownian motion process.

[31] Martingales and their properties. Doob's inequality for Martingales, the Martingale downcrossing inequality and the Martingale convergence theorem.

[32] Stochastic integration w.r.t a Martingale.

[33] Examples of Martingales.

[34] An introduction to large deviation theory with applications to the diffusion exit problem and stabilization of stochastic differential equations with feedback controllers.

[35] Stochastic processes in robotics:

[a] The d-link robot equation.

[b] The d-link robot equation with 3-D links–analysis using Lie group theory.

[c] Large deviation control of 3-D link robots.

[35] Markov chains and the Chapman-Kolmogorov equations. Examples including the pure birth process, the birth-death process, the telegraph process. The stationary distribution of a Markov chain.

[36] Derivation of the Fokker-Planck equations for a continuous state space Markov process from Ito's stochastic differential equation.

[37] Approximation of the Boltzmann kinetic transport equation for a plasma by the Fokker-Planck equation.

[38] An introduction to probability in quantum mechanics.

[a] Interference of wave functions.

[b] Interpretation of quantum probabilities using Feynman's path integral formula for the probability amplitude.

[c] Transition probabilities in quantum mechanics.

[d] Transition probabilities when the system Hamiltonian is perturbed by a time varying Hamiltonian-Development of time dependent perturbation theory.

[e] The quantum mechanical harmonic oscillator and its application to the construction of the Boson Fock space.

[f] The creation, annihilation and conservation processes of Hudson and Parthasarathy in Boson Fock space.

[g] Quantum stochastic integration and quantum stochastic differential equations in hte Hudson-Parthasarathy formalism for describing the evolution of quantum systems in the presence of quantum noise.

References:

[1] A.Papoulis, "Probability Theory, Random Variables and Stochastic Processes".

[2] William Feller, "An introduction to probability theory and its applications, vol.I and II", John Wiley.

[3] K.R.Parthasarathy, "An introduction to probability and measure", Hindustan Book Agency.

[4] K.R.Parthasarathy, "An introduction to quantum stochastic calculus", Birkhauser, 1992.

[5] Harish Parthasarathy, "Developments in Mathematical and Conceptual Physics:Concepts and Applications for Engineers", Springer Nature, 2020.

[6] I.Karatzas and S.Shreve, "Brownian motion and stochastic calculus", Springer.

2.1 The basic axioms of Kolmogorov

A triplet $(\Omega, \mathcal{F}, P)$ is called a classical probability space for an experiment where Ω is the sample space, namely a set whose elements are called elementary outcomes of the experiment, $\mathcal{F}$ is a σ-field of subsets of Ω and $P : \mathcal{F} \to [0,1]$ is a probability measure. By a σ-field, we mean that it is closed under countable unions and complementation (and hence also under countable intersections (by De-Morgan's rule $\bigcap_n E_n = (\bigcup_n E_n^c)^c$), and therefore it also contains the sample space as well as the nullset. The elements of $\mathcal{F}$ (which are subsets of Ω) are called the events of the experiment. If $E \in \mathcal{F}$, we say that the event E has occurred if on performing the experiment, the elementary outcome $\omega \in E$. We say that the event E has not occurred, ie E^c has occurred if the elementary outcome $\omega \in E^c$, or equivalently $\omega \notin E$. If $E_n, n = 1, 2, ...$ is a finite or infinite sequence of events, then the finite/countable union $\bigcup_n E_n$ is an event by hypothesis and this event is said to have occurred if the elementary outcome ω is in at least one of the $E_n's$. Likewise, if ω is in all the $E_n's$, ie $\omega \in \bigcap_n E_n$, then we say that all the events $E_n, n = 1, 2, ...$ have occurred. Note that $\mathcal{F}$ need not be closed under arbitrary unions/intersections. The reason for this is seen when we define the probability measure P as a countably additive set function on $\mathcal{F}$ (which means that $E = \bigcup_n E_n, E_n \cap E_m = \phi \forall n \neq m$ imply $P(E) = \sum_n P(E_n)$) such that $P(\Omega) = 1$ and hence $P(\phi) = 0$. Now suppose we assumed that $\mathcal{F}$ is closed under arbitrary unions, not necessarily countable and then we also make P additive under uncountable unions of disjoint events. Then, we run into trouble as the following example shows: Let P be the uniform distribution on the closed interval $[0,1]$. Then $P([0,1]) = 1$. However $P(\{x\}) = 0$ for any single point $x \in [0,1]$ for P by definition is given by $P([a,b]) = b - a$ for $0 \leq a \leq b \leq 1$. On the other hand, uncountable additivity of P would result in

$$1 = P([0,1]) = \sum_{x \in [0,1]} P(\{x\}) = \sum_{x \in [0,1]} 0 = 0$$

which is absurd. That is the reason why we have to be content with $\mathcal{F}$ being closed under countable unions and P being countably additive on $\mathcal{F}$.

2.2 Exercises

[1] Show that if $A, B \in \mathcal{F}$ and $A \subset B$, then

$$P(A) \leq P(B)$$

[2] Show using countable additivity of P on $\mathcal{F}$ that if $E_n \in \mathcal{F}, n = 1, 2, ...$ and $E_n \uparrow E$, ie, $E_n \subset E_{n+1} \forall n \geq 1$ and $\bigcup_{n \geq 1} E_n = E$, then

$$P(E_n) \uparrow P(E)$$

Conversely, show that if P is finitely additive on $\mathcal{F}$ and this property holds then P is countably additive on $\mathcal{F}$.

hint: Let $E_0 = \phi$ and define

$$F_n = E_n - E_{n-1} = E_n \cap E_{n-1}^c, n \geq 1$$

then the $F_n's$ are pairwise disjoint events and

$$\bigcup_n F_n = \bigcup_n E_n = E$$

Apply now the countable additivity property of P and use the fact that

$$P(F_n) = P(E_n) - P(E_{n-1})$$

[3] Show that if $E_n \downarrow E$ (all being events), ie, $E_{n+1} \subset E_n \forall n$ and $E = \bigcap_n E_n$, then

$$P(E_n) \downarrow P(E)$$

hint: $E_n \downarrow E$ iff $E_n^c \uparrow E^c$. Now use the result of the previous exercise. Conversely show that if P is finitely additive on $\mathcal{F}$ and this property holds good, then P is countably additive.

[4] Study project on the Caratheodory extension theorem. Let $\mathcal{B}$ field, ie, a collection of Ω-subsets that is closed under finite unions and complementation and if P is a countably additive probability measure on $\mathcal{B}$ (ie, $E_n \in \mathcal{B}, n \geq 1$, $E_n \cap E_m = \phi \forall n \neq m$ and $E = \bigcup_n E_n \in \mathcal{B}$, then $P(E) = \sum_n P(E_n)$), then P has a unique countably additive extension to a probability measure P_0 on the σ-field $\mathcal{F} = \sigma(\mathcal{B})$ generated by $\mathcal{B}$. By unique countable extension, we mean that (a) P_0 is a probability measure on $\mathcal{F}$ and (b) $P_0(E) = P(E) \forall E \in \mathcal{B}$

[5] If X_n is a bounded sequence of random variables such that $X_n \leq X_{n+1} \forall n$, then X_n increases to a limit X. Show that for X to be measurable, ie, a random variable in general, we require $\mathcal{F}$ to be a σ-field just being a field will not suffice.

hint: The set $\{\omega : X(\omega) \in (a, b]\}$ is the increasing limit of the events $\{\omega : X_n(\omega) \in (a, b]\}, n = 1, 2, ...$, in particular the former is the countable union of the latter. Hence for the former to be measurable, ie, an event, the class $\mathcal{F}$ of events must be closed under countable unions. Further, if we require the continuity condition

$$P(X \in (a, b]) = lim_n P(X_n \in (a, b])$$

then P must be countably additive in general, finite additivity will not suffice.

[6] Let $(\Omega_k, \mathcal{F}_k, P_k), k = 1, 2, ..., r$ be probability spaces. Let

$$\Omega = \Omega_1 \times \Omega_2 \times ... \times \Omega_r$$

and let $\mathcal{F}$ be the σ field on Ω generated by the measurable rectangles, ie, by sets of the form $E_1 \times E_2 \times .... \times E_r$ with $E_m \in \mathcal{F}_m, m = 1, 2, ..., r$. Let $\mathcal{B}$ denote the field consisting of finite disjoint unions of such rectangles (Show that this is indeed a field). It is clear that $\mathcal{F}$ is the σ-field generated by $\mathcal{B}$. Prove using the countable additivity of P_k on $\mathcal{F}_k, k = 1, 2, ..., r$ that the finitely additive set function P defined on $\mathcal{B}$ by

$$P(A_1 \cup A_2 \cup ... \cup A_r) = P(A_1) + ... + P(A_r)$$

where $A_1, ..., A_r$ are disjoint measurable rectangles and if $A = E_1 \times .... \times E_r$ is a measurable rectangle, then

$$P(A) = P_1(E_1)...P_r(E_r)$$

is also countably additive on $\mathcal{B}$ and hence use Caratheodory's extension theorem to deduce that P extends to a unique probability measure P_0 on $\mathcal{F}$ (When we say probability measure, we mean that it should be countably additive).

Note: In order to show that P is countably additive on $\mathcal{B}$, it suffices to show that if $E_n \in \mathcal{B}, E_n \downarrow \phi$ then $P(E_n) \downarrow 0$.

[7] The Kolmogorov existence theorem for stochastic processes. Let $F_n(x_1, ..., x_n), n \geq 1$ be a consistent family of probability distributions on $\mathbb{R}^n, n = 1, 2, ...$ respectively. Then, if

$$\mathcal{B} = \bigcup_{n \geq 1} (\mathcal{B}(\mathbb{R}^n) \times \mathbb{R}^{\mathbb{Z}_+})$$

prove that $\mathcal{B}$ is a field in $\mathbb{R}^{\mathbb{Z}_+}$.

Exercises:

[1] if $X_1, ..., X_n, ...$ is a sequence of random variables on a probability space $(\Omega, \mathcal{F}, P)$, then show that if we define the joint probability distribution function of the first n r.v's by

$$F_n(x_1, ..., x_n) = P(X_1 \leq x_1, ..., X_n \leq x_n) = P(\bigcap_{k=1}^{n} X_k^{-1}((-\infty, x_k])), x_1, ..., x_n \in \mathbb{R}$$

then $F_n, n \geq 1$ has the following properties:

$$lim \downarrow x_i y_i F_n(x_1, .., x_i, .., x_n) = F_n(x_1, ..., y_i, ..., x_n)$$

ie, F_n is right continuous in each of its arguments. To prove this, make use of the fact that $[-\infty, y_i) = lim x_i \downarrow y_i(-\infty, x_i]$ and hence $lim x_i \downarrow y_i X_i^{-1}((-\infty, x_i]) = X_i^{-1}((-\infty, y_i])$. Hence, deduce that

$$lim x_i \downarrow y_i \bigcap_{k=1}^{n} X_k^{-1}((-\infty, x_k]) = X_1^{-1}((-\infty, x_1]) \bigcap ... \bigcap X_i^{-1}((-\infty, y_i]) \bigcap ... \bigcap X_n^{-1}((-\infty, x_n])$$

Then, make use of the continuity of the probability measure P (which is a consequence of the countable its additivity) to deduce the result.

2.3 Exercises on stationary stochastic processes, spectra and polyspectra

[1] Let $X(t), t \in \mathbb{R}$ ($X(n), \in \mathbb{Z}$) be a stochastic process in continuous time (discrete time). The process is said to be stationary if the joint distribution of the random variables $(X(t), X(t+t_1), ..., X(t+t_k))$ does not depend on t for any $k, t_1, ..., t_k$. In this case, define the $(k+1)^{th}$ order moments of the process as

$$M_X(t_1, ..., t_k) = \mathbb{E}(X(t)X(t+t_1)...X(t+t_k))$$

Show that this does not depend upon t, ie, the process is $(k+1)^{th}$-order stationary. Second order stationarity in particular means that the autocorrelation function $R_X(s) = \mathbb{E}(X(t)X(t+s))$ does not depend on t. Give an example of a process that is second order stationary but is not stationary.

Then, define its k-variate Fourier transform by

$$P_{X,k}(\omega_1, ..., \omega_k) = \int_{\mathbb{R}^k} M_X(t_1, ..., t_k) exp(-j(\omega_1 t_1 + ... + \omega_k t_k)) dt_1...dt_k$$

$P_{X,k}$ is called the kth order polyspectrum of the process X. In the discrete time case, we define it using the k-variate DTFT of the moment sequence rather than the continuous time FT, ie, CTFT.

Show that if $X(t)$ is passed through an LTI system with impulse response $h(t)$ so that its output is

$$Y(t) = \int_{\mathbb{R}} h(s)X(t-s)ds$$

or in discrete time,

$$Y(n) = \sum_{m \in \mathbb{Z}} h(m)X(n-m)$$

then

$$P_{Y,k}(\omega_1, ..., \omega_k) = H(\omega_1)...H(\omega_k)\bar{H}(\omega_1 + ... + \omega_k)P_{X,k}(\omega_1, ..., \omega_k), k = 1, 2, ...$$

In particular, show that

$$S_Y(\omega) = P_{Y,2}(\omega) = |H(\omega)|^2 S(X(\omega), S_X(\omega) = P_{X,2}(\omega)$$

A process that is both first and second order stationary is said to be wide sense stationary (WSS). Let $X(t)$ be a WSS process and define its time and ensemble average power by

$$W_X = lim_{T \to \infty} \mathbb{E} \frac{1}{T} \int_{-T/2}^{T/2} X(t)^2 dt$$

Prove using the Parseval theorem that

$$W_X = \frac{1}{2\pi}\int_{\mathbb{R}} S_X(\omega)d\omega$$

where

$$S_X(\omega) = lim_{T\to\infty}\frac{1}{T}\mathbb{E}|\hat{X}_T(\omega)|^2 = \int_{\mathbb{R}} R_X(s)exp(-j\omega s)ds$$

where

$$\hat{X}_T(\omega) = \int_{-T/2}^{T/2} X(t)exp(-j\omega t)dt$$

and

$$R_X(s) = \mathbb{E}(X(t)X(t+s))$$

For this reason $S_X(\omega)$ is called the power spectral density (PSD) of the WSS process $\{X(t) : t \in \mathbb{R}\}$. The above result, namely that the PSD of a WSS process is the Fourier transform of its autocorrelation function is called the Wiener-Khintchine theorem.

For proving this, you must assume that

$$lim_{|s|\to\infty}R_X(s) = 0$$

which in particular, is true if

$$\int_{\mathbb{R}} |R_X(s)|ds < \infty$$

2.4 A research problem based on problem [1]

Explain how using measurements of the power spectral density of the input and output of an LTI system, you can estimate the magnitude $|H(\omega)|$ of the transfer function of the system and by using measurements of the polyspectrum of order k where $k \geq 3$ of the input and outputs of the LTI system, we can also estimate the phase of the LTI system.

[3] Let $Z(\omega), \omega \in \mathbb{R}$ be a zero mean complex valued stochastic process on a probability space such that

$$\mathbb{E}(dZ(\omega).d\bar{Z}(\omega')) = (2\pi)^{-1}S(\omega)d\omega.\delta_{\omega,\omega'}$$

Now define a stochastic process $X(t)$ as the stochastic integral

$$X(t) = \int_{\mathbb{R}} exp(j\omega t)dZ(\omega)$$

Show that $X(t)$ is WSS with autocorrelation

$$R_X(s) = \mathbb{E}(X(t+s)\bar{X}(t)) = \int_{\mathbb{R}} exp(j\omega s)S(\omega)d\omega/2\pi$$

Hence deduce that

$$S(\omega) = 2\pi.\mathbb{E}(|dZ(\omega)|^2)/d\omega$$

is the power spectral density of the process $X(t)$. Conversely, given a WSS process $X(t)$, define a process $Z(\omega), \omega \in \mathbb{R}$ by the equation

$$Z(\omega_2)-Z(\omega_1) = (2\pi)^{-1}\int_{\mathbb{R}} \frac{(exp(-j\omega_2 t) - exp(-j\omega_1 t))}{-jt} X(t)dt, -\infty < \omega_1 < \omega_2 < \infty$$

Then show that formally we can write

$$dZ(\omega)/d\omega = (2\pi)^{-1}\int_{\mathbb{R}} exp(-j\omega t)X(t)dt, \omega \in \mathbb{R}$$

Note that when we are rigorous, this statement is true only if the complex measure on $\mathbb{R}$ defined by

$$\mu_Z((\omega_1, \omega_2]) = Z(\omega_2) - Z(\omega_1)$$

is absolutely continuous w.r.t the Lebesgue measure. Show that even if it is not so, but the limit

$$lim_{\omega_2 \to \omega_1}\mathbb{E}(|Z(\omega_2) - Z(\omega_1)|^2)/(\omega_2 - \omega_1)$$

$$= (2\pi)^{-1}S(\omega_1)$$

exists for all $\omega_1 \in \mathbb{R}$, then show also by virtue of the WSS property of $X(t)$ that we have the orthogonality relations

$$\mathbb{E}[(Z(\omega_2) - Z(\omega_1).(\bar{Z}(\omega_3) - \bar{Z}(\omega_4))] = 0$$

for

$$\omega_2 > \omega_1 \geq \omega_3 > \omega_4$$

and hence deduce the relation

$$R_X(s) = \mathbb{E}(X(t+s)\bar{X}(t)) = (2\pi)^{-1}\int_{\mathbb{R}} S(\omega)exp(j\omega s)d\omega$$

in the sense of Riemann-Stieltjes. In this case, show that we can define the integral

$$\int_{\mathbb{R}} exp(j\omega t)dZ(\omega)$$

in the L^2 sense as an L^2-limit of Riemann sums and that this L^2 limit equals $X(t)$.

[4] This problem is a generalization of the previous problem. Let $(\Omega, \mathcal{F}, P)$ be a probability space and let $\mathcal{H} = L^2(\Omega, \mathcal{F}, P)$ denote the Hilbert space of all complex valued random variables X on this probability space for which

$$\mathbb{E}|X|^2 = \int |X(\omega)|^2 dP(\omega) < \infty$$

Let Z be a complex set function on a measurable space $(X, \mathcal{E})$ with the property that Z is countably additive in the L^2-sense, ie, if $E_1, E_2, ...$ is a sequence of pairwise disjoint sets in $\mathcal{E}$, then

$$\mathbb{E}|\mu(\bigcup_n E_n) - \sum_{n=1}^{N} \mu(E_n)|^2 \to 0, N \to \infty - - - (1)$$

Let μ be a measure on $(X, \mathcal{E})$, ie $(X, \mathcal{E}, \mu)$ is a measure space. Assume that

$$\mathbb{E}(Z(A).\bar{Z}(B)) = \mu(A \cap B), A, B \in \mathcal{E}$$

Show using the countable additivity of μ on $\mathcal{E}$, that this condition automatically guarantees that $Z(.)$ will be countably additive in the L^2-sense, ie, the property (1) will hold.

Let f be a complex valued measurable function on this measure space and assume that

$$\int_X |f(x)|^2 d\mu(x) < \infty$$

ie,

$$f \in L^2(X, \mathcal{E}, \mu)$$

We wish to define a stochastic integral

$$\int_X f(x) dZ(x) \in \mathcal{H}$$

in the L^2 sense and elucidate some properties of this stochastic integral. Choose a simple sequence of measurable functions f_n on $(X, \mathcal{E}, \mu)$ converging in the L^2 sense to f, ie, each f_n has the form

$$f_n(x) = \sum_{k=1}^{N_n} c(n,k) \chi_{E_{n,k}}(x), n \geq 1$$

By saying that this sequence converges to f in the L^2-sense, we mean that

$$\int_X |f_n(x) - f(x)|^2 d\mu(x) \to 0, n \to \infty$$

Then define

$$I_Z(f_n) = \sum_{k=1}^{N_n} c(n,k) Z(E_{n,k})$$

Show that $\{I_Z(f_n)\}$ is a Cauchy sequence in $\mathcal{H}$, or more precisely,

$$\mathbb{E}|I_Z(f_n) - I_Z(f_m)|^2 = \int_X |f_n(x) - f_m(x)^2 d\mu(x) \to 0, n, m \to \infty$$

where the last convergence follows from the fact that every convergent sequence in a Hilbert space (or more generally, in any inner product space) is Cauchy. Deduce that there exists an element $I_Z(f) \in \mathcal{H}$ such that

$$\mathbb{E}|I_Z(f_n) - I_Z(f)|^2 \to 0$$

and that this L^2 limit $I_Z(f)$ does not depend upon the sequence f_n of simple functions converging to f. We write

$$I_Z(f) = \int_X f(x) dZ(x)$$

and call it the L^2-stochastic integral of f w.r.t Z.

2.5 Exercises on the construction of the integral w.r.t a probability measure

[1]

[a] Let $(\Omega, \mathcal{F}, P)$ be a probability space and let X be a random variable on it. We say that X is a simple r.v. if it assumes atmost only a finite number of distinct values, say $c_1, ..., c_n$. Define

$$E_k = X^{-'1}(\{c_k\}) = \{\omega \in \Omega : X(\omega) = c_k\}. k = 1, 2, ...n$$

Show that $E_1, ..., E_n$ are disjoint events, ie,

$$E_k \cap E_j = \phi, k \neq k, E_k \in \mathcal{F}$$

and further

$$\Omega = \bigcup_{k=1}^{n} E_k$$

Show that we can write

$$X(\omega) = \sum_{k=1}^{n} c_k \chi_{E_k}(\omega)$$

Define

$$\int X dP = \sum_{k=1}^{n} c_k P(E_k)$$

Show that if we write the same r.v X in another way as

$$X(\omega) = \sum_{k=1}^{m} d_k \chi_{F_k}(\omega)$$

where the $F_k^{\prime}s$ need not be disjoint (but they are events), then

$$\sum_{k=1}^{m} d_k P(F_k) = \int X dP = \sum_{k=1}^{n} c_k P(E_k)$$

[b] Show that the set of simple r.v.s is a vector space over the real number, ie, it is closed under addition and scalar multiplication by real numbers. Therefore, this set is also closed under all finite real linear combinations.

[c] Show that if X, Y are simple r.v's and $X(\omega) \leq Y(\omega) \forall \omega \in \Omega$, then

$$\int X dP \leq \int Y dP$$

In particular, deduce using [b] that if X is a non-negative simple r.v. then

$$\int X dP \geq 0$$

[2] First we construct the integral of a non-negative r.v. Let $(\Omega, \mathcal{F}, P)$ be a probability space and X a non-negative real valued random variable on this space. A simple r.v. is a r.v. that assumes only a finite number of values. For each positive integer N, define the simple r.v X_N by

$$X_N(\omega) = \sum_{k=0}^{N.2^N} (k/2^N) \chi_{X^{-1}((k/2^N,(k+1)/2^N])}(\omega)$$

where $\chi_E(\omega)$ denotes the indicator of E, ie, $\chi_E(\omega) = 1$ if $\omega \in E$ and $\chi_E(\omega) = 0$ if $\omega \notin E$. Using the decomposition

$$(k/2^N, (k+1)/2^N] = (2k/2^{N+1}, (2k+1)/2^{N+1}] \cup ((2k+1)/2^{N+1}, (2k+2)/2^{N+1}]$$

of the lhs into a disjoint union, deduce that

$$0 \leq X_N(\omega) \leq X_{N+1}(\omega) \forall N \geq 1$$

ie, $X_N, N \geq 1$ is a non-decreasing sequence of simple r.v.s Show further that

$$lim_{N\to\infty} X_N(\omega) = X(\omega) \forall \omega \in \Omega$$

Deduce using the result of the previous exercise that

$$0 \leq \int X_N dP \leq \int X_{N+1} dP < \infty, \forall N \geq 1$$

and hence that

$$I = lim_{N\to\infty} \int X_N dP$$

Exists. Define

$$\int XdP = I$$

Let $Y_N, N \geq 1$

2.6 Exercises on stationarity, dynamical systems and ergodic theory

[1] Let $f \in L^1(\Omega, \mathcal{F}, P)$ and let $T : \Omega \to \Omega$ be a measure preserving transformation, ie,

$$T^{-1}(\mathcal{F}) \subset \mathcal{F}, PoT^{-1} = P$$

if T is invertible and X is a random variable on $(\Omega, \mathcal{F}, P)$, then show that the process $X(T^n\omega), n \in \mathbb{Z}$ is a stationary stochastic process on $(\Omega, \mathcal{F}, P)$.

Chapter 3

Antenna Theory

3.1 Course Outline

[1] Maxwell's equations in the frequency domain. Solution using retarded potentials, the far field approximation. Calculation of the total power radiated at a given frequency in the far field zone.

[2] Maxwell's equations in the frequency domain taking inhomogeneities, anisotropicity and field dependence (nonlinearity) of the permittivity and permeability tensors into consideration. Perturbative solution of the differential equations.

[3] Construction of the Green's function of the Helmholtz operator for boundaries of various shapes for the Dirichlet and Neumann boundary conditions:Applications to cavity resonator antennas.

[4] The reciprocity theorem between two electromagnetic fields driven by two different current densities.

[5] The radar equation, directivity and antenna aperture, reciprocity theorem between transmitter and receiver antenna.

[6a] The basic antenna parameters: Loss resistance, radiation resistance total resistance, effective aperture. Equivalent circuit of a transmitter and receiver antenna.

[6b] Poynting's theorem and its evaluation in the far field zone in terms of the current distribution in the source.

[7] The fields of an infinitesimal electric dipole in the near field zone, far field zone, intermediate zone.

[8] The fields produced by a finite straight wire of length L carrying a spatial sinusoid current distribution vanishing at its ends.

[9] The fields produced by a circular loop carrying sinusoidal current in the far field zone.

[10] The far field pattern produced by an infinitesimal loop of wire in terms of the magnetic moment of the wire.

[11] The far field pattern produced by a volume carrying a sinusoidal current density.

[12] The far field pattern produced by an infinitesimal dipole, a finite straight wire and a circular loop placed above an infinite ground plane. Calculation based on the method of images.

[13] The total power radiated by an antenna in the far field zone.

[14] The radiation resistance of an antenna in terms of the current distribution in it and the feeding current.

[15] A planar aperture on which an electromagnetic field is incident as an antenna. Computation of the far field Poynting vector.

[16] A waveguide terminated by a horn:The far field radiation pattern.

[17] Helical antennas as broad band antennas.

[18] A cavity resonator as a microstrip antenna.

[19] Waveguide feeding an aperture as an antenna.

[20] The method of stationary phase.

[21] The mutual impedance between two antennas, the self impedance of an antenna.

[22] Antenna arrays: The pattern multiplication theorem, application to binomial and Chebyshev arrays.

[23] Plotting of antenna pattern lobes.

[24a] Computing the current density induced on an antenna surface by an incident electromagnetic field using generalizations of the Pocklington and Hallen integral equations.

[24b] Application to the problem of determining the induced currents on the driver and driven elements of a Yagi-Uda array.

[24c] Solving Pocklington's integral equations by the method of moments.

[25] The effect of a gravitational field and inhomogeneous, anisotropic and field dependent permittivity and permeability on the pattern of an antenna:General relativistic considerations based on Covariant form of the Maxwell equations in a background curved space-time.

[26] A brief description of the finite element method for numerically solving antenna problems.

[25] An introduction to quantum antennas.

[a] Quantum current generated by electrons, positrons and photons within a cavity on the atomic scale. Description of currents on the cavity surface generated by the quantum magnetic field and the currents within the cavity generated by the electrons-positron Dirac field. Evaluation of the quantum statistical moments of the near and far field radiation patterns in a given state of the electrons, positron and photons. A description of coherent states for photons and electrons-positrons.

3.2 The far field Poynting vector

The far field magnetic vector potential is given by

$$\mathbf{A}(t,\mathbf{r}) = (\mu/4\pi)\int \mathbf{J}(\omega,\mathbf{r})exp(-jk|\mathbf{R}-\mathbf{r}|)d^3r/|\mathbf{R}-\mathbf{r}|$$

$$\approx \mathbf{P}(\omega,\hat{R})exp(-jkR)/R$$

where

$$\mathbf{P}(\omega,\hat{R}) = (\mu/4\pi)\int_S \mathbf{J}(\omega,\mathbf{r})exp(jk\hat{R}.\mathbf{r})d^3r$$

where S denotes the source volume. Thus, the far field electric field (ie upto $O(1/R)$) is given by

$$\mathbf{E}(\omega,\mathbf{R}) = -\nabla\Phi - j\omega\mathbf{A}$$

$$= -\nabla((jc^2/\omega)div\mathbf{A})) - j\omega\mathbf{A}$$

$$= [-(jc^2/\omega)(-jk\hat{R}.\mathbf{P})(-jk\hat{R})/R - j\omega\mathbf{P}/R]exp(-jkR)/R$$

$$= [(jk^2c^2/\omega)P_r\hat{R}/R - j\omega(P_r\hat{R} + P_\theta\hat{\theta} + P_\phi\hat{\phi})]exp(-jkR)/R$$

$$= -j\omega(P_\theta\hat{\theta} + P_\phi\hat{\phi})exp(-jkR)/R$$

Note: The only $O(1/R)$ term in when we take the gradient of a function of the form $f(R,\theta,\phi)exp(-jkR)/R$ is given by

$$-jk\hat{R}.f(R,\theta,\phi)/R$$

All the other terms are of order $1/R^2$ and they do not contribute anything to the outward radiated power. In other words, the $O(1/R)$ terms come only by differentiation of the phase factor or equivalently from differentiation of the far field delay factor, not by differentiating the amplitude factor. We can likewise evaluate the far field magnetic field (ie, with neglect of $O(1/R^2)$ terms):

$$\mathbf{B}(\omega,\mathbf{R}) = \nabla\times\mathbf{A}(\omega,\mathbf{R}) =$$

$$-jk\hat{R}\times\mathbf{P}.exp(-jkR)/R$$

$$= -jk(P_\theta\hat{\phi} - P_\phi\hat{\theta}).exp(-jkR)/R$$

Therefore, the time averaged far field Poynting vector, ie, power flux is given by (upto $O(1/R^2)$ terms)

$$\mathbf{S}(\omega,R,\hat{R}) = \mathbf{S}(\omega,R,\theta,\phi) = Re[\mathbf{E}\times\mathbf{B}^*]/2\mu_0 =$$

$$= [\omega k(|P_\theta|^2 + |P_\phi|^2)/2\mu_0R^2]\hat{R}$$

$$= (\omega^2/2\mu_0cR^2)(|P_\theta(\omega,\hat{R})|^2 + P_\phi(\omega,\hat{R})|^2)\hat{R}$$

where we use the relations

$$\hat{\theta}\times\hat{\phi} = \hat{R}$$

The total power radiated out by the antenna when it operates at the frequency ω is then

$$W = \int_{sphereorradiusR} \mathbf{S}(\omega, R, \hat{R}).\hat{R}dS(\hat{R})$$

$$= \int_{unitsphere} \mathbf{S}(\omega, R, \hat{R}).\hat{R}.R^2 d\Omega(\hat{R})$$

$$= (\omega^2/2\mu_0 c)\int_0^{\pi}\int_0^{2\pi} |P_\theta(\omega,\theta,\phi)|^2 + |P_\phi(\omega,\theta,\phi)|^2)sin(\theta)d\theta.d\phi$$

Note that this result is independent of the radial distance R as long as we are operating in the far field zone. This formula can be used to define the radiation resistance R_r of the antenna at frequency ω as

$$|I_0(\omega)|^2 R_r/2 = W$$

where $I_0(\omega)$ is the input sinusoidal current phasor at frequency ω.

Reciprocity theorem: Sources $(\rho_1, J_1, \rho_{m1}; M_1)$ generate fields (E_1, H_1) while sources $(\rho_2, J_2, \rho_{m2}, M_2)$ generate fields (E_2, H_2). Thus,

$$curlE_k = -j\omega\mu H_k - M_k, curlH_k = j\omega\epsilon E_k + J_k, k = 1, 2$$

Compute

$$div(E_1 \times H_2) = curl(E_1).H_2 - E_1.curl(H_2) =$$

$$= (-j\omega\mu H_1 - M_1).H_2 - E_1.(j\omega\epsilon E_2 + J_2),$$

$$div(E_2 \times H_1) = curl(E_2).H_1 - E_2.curl(H_1) =$$

$$= (-j\omega\mu H_2 - M_2).H_1 - E_2.(j\omega\epsilon E_1 + J_1),$$

Thus,

$$div(E_1 \times H_2 - E_2 \times H_1) = M_2.H_1 - M_1.H_2 + J_1.E_2 - J_2.E_1$$

Integrating this identity over the entire three dimensional space, making use of Gauss' integral theorem and the fact that the electromagnetic fields vanish at ∞ gives us the fundamental reciprocity relation

$$\int_{\mathbb{R}^3} (J_1.E_2 - J_2.E_1 + M_2.H_1 - M_1.H_2)d^3r = 0$$

3.3 Exercises

3.3.1 Equivalent circuit of a transmitter-receiver antenna system

3.3.2 Radiation resistance of an infinitesimal dipole starting from the relation

$$\mathbf{A}(\omega, r) = (\mu dl/4\pi)\hat{z}.exp(-jkr)/r$$

While calculating the fields by differentiating the potentials, make use of the fact that spatial derivatives of only the phase term $exp(-jkr)$ need to be taken not of the other amplitude terms, for only the phase derivatives contribute to $O(1/r)$ terms in the field and hence to $O(1/r^2)$ terms in the Poynting vector in the far field zone and hence to non-zero total radiated power. Amplitude derivatives of the potentials contribute to $O(1/r^2)$ terms in the fields and hence to $O(1/r^3)$ terms in the Poynting vector which do not contribute anything to the radiated power in the far field zone.

Note that the electric field can be computed using

$$j\omega\epsilon E = curlB = curlcurlA = \nabla(divA) - \nabla^2 A = \nabla(divA) + k^2 A$$

and the magnetic field using

$$H = curlA/\mu$$

Thus, upto $O(1/r)$, we have

$$H = curlA/\mu = (-jk.dl/4\pi)\hat{r} \times \hat{z}.exp(-jkr)/r$$

$$= (jkdl/4\pi r)sin(\theta)\hat{\phi}.exp(-jkr)$$

and hence also

$$E = curlH/j\omega\epsilon = (-jk^2 dl/(\omega\epsilon 4\pi r)sin(\theta)\hat{r} \times \hat{\phi}.exp(-jkr)$$

$$= (jk^2 dl/\omega\epsilon.4\pi r)sin(\theta)\hat{\theta}.exp(-jkr)$$

where we have used

$$\hat{r} \times \hat{z} = -sin(\theta)\hat{\phi},$$

$$\hat{r} \times \hat{\phi} = -\hat{\theta}$$

Note that the above formulae imply

$$E_\theta/H_\phi = k/\omega\epsilon = 1/c\epsilon = \sqrt{\mu/\epsilon}$$

which is the characteristic impedance of the medium, as it should be.

[3] Show that for a small loop of wire carrying a current $I(\omega)$ located around the origin of the coordinate system with the parametric equation of the loop being given by $s \to \mathbf{R}(s), 0 \leq s \leq 1$, the far field magnetic vector potential is given by

$$\mathbf{A}(\omega, \mathbf{r}) = (\mu I(\omega)/4\pi) \int_0^1 d\mathbf{R}(s).exp(-jk|\mathbf{r} - \mathbf{R}(s)|)/|\mathbf{r} - \mathbf{R}(s)|$$

$$\approx (\mu I/4\pi r)exp(-jkr). \int_0^1 d\mathbf{R}(s)exp(j\hat{r}.\mathbf{R}(s))$$

Show that when this is a circular loop of radius a, the above formula simplifies to

$$\mathbf{A}(\omega, r, \theta, \phi) =$$

$$(\mu Ia/4\pi r)exp(-jkr) \int_0^{2\pi} (-\hat{x}.sin(\phi')+\hat{y}.cos(\phi')).exp(jka.sin(\theta)cos(\phi'-\phi))d\phi'$$

and then using the formulas

$$A_\rho = A_x.cos(\phi) + A_y.sin(\phi), A_\phi = -A_x.sin(\phi) + A_y.cos(\phi)$$

deduce that in the far field zone.

$$A_\rho = 0, A_z = 0,$$

$$A_\phi = (\mu Ia/4\pi r)exp(-jkr) \int_0^{2\pi} exp(jka.sin(\theta).cos(\phi'))cos(\phi')d\phi'$$

Problem: From the expression for the far field electromagnetic field pattern upto $O(1/r^2)$ for a source at definite frequency, calculate the Poynting vector field upto $O(1/r^2)$ and hence determine the total power radiated out into space in the far field zone as a quadratic form in the source current density. Specialize this result to infinitesimal dipoles and to finite length dipoles and circular loops carrying sinusoidal current and hence calculate the radiation resistance R using $P = < I^2 > R/2$.

Appendix, B.E and M.Tech projects

3.4 Order of magnitudes in quantum antenna theory

Consider a cavity resonator of one Angstrom size, ie, a cube with each side of length $a = 10^{-10}m$. The Maxwell equations in such a cube have solutions of the from

$$A_r(t, x, y, z) = \sum_{mnp} c(mnp, t)u_{r,mnp}(x, y, z), r = 1, 2, 3$$

where $u_{r,mnp}$ are spatial functions obtained by integrating the electric field w.r.t time. These functions are of the form

$$\{cos(m\pi x/a), sin(m\pi x/a)\} \otimes \{cos(n\pi y/a), sin(n\pi y/a)\} \otimes \{cos(p\pi z/a), sin(p\pi z/a)\}$$

multiplied by some constants depending on the indices (m, n, p). We may, without loss of generality, assume that the functions $u_{r,mnp}$ are normalized so that

$$\int_C u_{r,mnp}(\mathbf{r})\bar{u}_{s,m'n'p'}(\mathbf{r})d^3r = \delta_{rs}\delta_{mm'}\delta_{nn'}\delta_{pp'}$$

The dependence of $c(mnp, t)$ on t is $exp(i\omega(mnp)t)$ where $\omega(mnp)$ are the characteristic frequencies of oscillation:

$$\omega(mnp) = (\pi c/a)\sqrt{m^2 + n^2 + p^2}, m, n, p = 1, 2, ...$$

which are of the order of magnitude

$$\omega = \pi c/a$$

The electric field is

$$E_r = \partial_t A_r = \sum_{mnp} c(mnp, t) i\omega(mnp) u_{r,mnp}(\mathbf{r})$$

The magnetic field is

$$B = curl A$$

which is of the order of magnitude $|c(mnp, t)|/a$ where by $c(mnp, t)$ we actuall mean its average in a coherent state. The total electric field energy within the cavity C is

$$U_E = (\epsilon_0/2) \int_C |E|^2 d^3r$$

which has components of the order of magnitude

$$\epsilon_0 |\omega(mnp)c(mnp, t)|^2 a^3 = \epsilon_0 \omega(mnp)^2 a^3 |c(mnp, t)|^2$$

The total magnetic field energy within the cavity is

$$U_B = (2\mu_0)^{-1} \int_C |B|^2 d^3r$$

which is has components of the order of magnitude

$$|c(mnp, t)/a|^2 a^3/\mu_0 = |c(mnp, t)|^2 a/\mu_0$$

The ration of the orders of magnitude of the electric field energy and the magnetic field energy within the cavity therefore has the order of magnitude

$$U_E/U_B \approx \mu_0 \epsilon_0 \omega(mnp)^2 a^2 \approx \omega^2 a^2/c^2 \approx 1$$

as expected. The canonical commutation relations are

$$[A_r(t,\mathbf{r}), \partial_t A_s(t,\mathbf{r}')] = (ih/2\pi)\delta^3(\mathbf{r}-\mathbf{r}')$$

These yield,

$$[c_r(mnp,t), \omega(mnp)c_s(m'n'p',t)^*] = (h/2\pi)\delta_{rs}\delta_{mm'}\delta_{nn'}\delta_{pp'}$$

so that the eigenvalues of $c_r(mnp,t)^*c_r(mnp,t)$ are positive integer multiples of $h/2\pi\omega(mnp)$. This means that the field energy within the cavity when a finite number of modes are excited assumes eigenvalues that are of the same order of magnitude as positive integer multiples of $h\omega/2\pi$ as expected by Planck's quantum theory of radiation. This fact also yields the result that $|c(mnp,t)|$ is of the order of magnitude of $\sqrt{h/(2\pi\omega)}$.

Now we come to the question of computing the order of magnitude of the Poynting vector power flux at a given radial distance R from the quantum cavity antenna caused by the surface current density induced by the magnetic field on on the antenna surface. The magnetic field on the surface and hence the corresponding induced surface current density both have the order of magnitudes of $|c(mnp,t)|/a$ which is of the order $a^{-1}\sqrt{h/\omega}$. Therefore, the far field magnetic vector potential at a distance R from the cavity is of the order of magnitude (use the retarded potential formula) $(a/R)\sqrt{h/\omega}$ and hence the corresponding far field radiated magnetic field is of the order of magnitude $(\omega/c)(a/R)\sqrt{h/\omega}$ while the near field magnetic field is of the order of magnitude $(a/R^2)\sqrt{h/\omega}$. Actually, these expressions for the magnetic field must be multiplied by $\sqrt{N}$ where N is a positive integer corresponding to the largest modal eigenvalue of the operators $(2\pi\omega(mnp)/h)c(mnp,t)^*c(mnp,t)$.

The far field Poynting vector has the order of magnitude of $B^2c/2\mu_0$ which is of the order

$$(c/2\mu_0)(\sqrt{N}.(\omega/c)(a/R).\sqrt{h/\omega})^2$$
$$= (h/2\mu_0)N.(\omega/c)(a^2/R^2)$$

and the total power radiated outward by this quantum antenna in the far field zone is thus of the order of magnitude

$$P = N(h/2\mu_0)(a^2\omega/c)$$

Now we look at the order of magnitude of the power radiated in the far field zone by the Dirac field of electrons and positrons within the cavity. The Dirac equation is

$$[i\gamma^\mu\partial_\mu - m]\psi(x) = 0$$

or more precisely in arbitrary units,

$$[(ih/2\pi)\partial_t - c(\alpha,(-ih/2\pi)\nabla) - \beta mc^2]\psi(x) = 0$$

Here, the appearance of the constants h, m, c is explicitly shown. Now the $|\psi(x)|^2$ is the probability density of the electron which must integrate to unity over the cavity volume. Thus $\psi(x)$ is of the order of magnitude $a^{-3/2}$. The Dirac current density $J^\mu = e\psi^*\gamma^0\gamma^\mu\psi$ has the same order of magnitude as $e|\psi(x)|^2c$ which is $ec/a^{3/2}$. Therefore the far field magnetic vector potential at a radial distance of R from the cavity is, in accordance with the retarded potential theory of the order

$$(ec/a^{3/2}).(a^3/R) = eca^{3/2}/R$$

The electric field in the far field zone is then of the order

$$E \approx \omega.eca^{3/2}/R$$

where ω is the characteristic oscilation frequency of the Dirac current. The magnetic field is of the order

$$B \approx a^{-1}.eca^{3/2}/R = ec\sqrt{a}/R$$

If P is the characteristic momentum of the electrons and positrons in a given state, for example P may be the average momentum of an electron in a given state, then according to De-Broglie, P is of the order h/a since a is the order of the electron wavelength. Then the electron energy is of the order

$$E_e = c\sqrt{m^2c^2 + P^2} \approx c\sqrt{m^2c^2 + h^2/a^2}$$

and the characteristic frequency of oscillation of the Dirac wave field is then

$$\omega = E_e/h$$

The Poynting vector corresponding to the power radiated by the Dirac field in the far field zone then has the order of magnitude

$$S \approx c(\epsilon_0 E^2 + B^2/\mu_0) = c^3\epsilon_0\omega^2 ea^3/R^2 + e^2c^3a/\mu_0 R^2$$

and the total power radiated in the far field zone is of the order

$$W = SR^2 = c^3\epsilon_0\omega^2 ea^3 + e^2c^3a/\mu_0$$

3.5 The notion of a Fermionic coherent state and its application to the computation of the quantum statistical moments of the quantum electromagentic field generated by electrons and positrons within a quantum antenna

Aim: The aim of this section is to present a calculation involving the computation of the quantum statistical moments of the electromagnetic field produced by an ensemble of electrons and positrons whose state is specified by a mixed state superposition of Fermionic coherent states. Fermionic coherent states are parameterized by Grassmannian/Fermionic numbers and in order to attach physical signficance to the final results, we must use the Berezin integral for Fermionic variables to determine the above mentioned superposition of Fermionic coherent states. We can incorporate some unknown real parameters into the Berezin linear combination of coherent states and estimate these parameters by minimizing the distance between the average value of the electromagnetic field generated by the Fermions and the desired electromagnetic field pattern. If need be, we may modify this cost function to be minimized by constraining the higher order quantum statistical moments of the generated quantum electromagnetic field to be specified. An example of an application of this circle of ideas is to use a quantum antenna to generate a set of desired spatial patterns at a given set of frequencies.

First consider just a single Fermion specified by the annihilation operator a and the creation operator a^*. Thus,

$$a^2 = a^{*2} = 0, aa^* + a^*a = 1$$

Let γ be a Grassmannian variable that will be used to specify the coherent state of this Fermion just as a complex number z is used to specify a the coherent state of a single Boson. γ anticommutes with itself, with γ^* and with a, a^*, just as in the Bosonic situation, the complex number z that specifies the coherent state commutes with itself, with $\bar{z}$ and with the Boson creation and annihilation operators:

$$\gamma^2 = 0, \gamma\gamma^* + \gamma^*\gamma - 0, \gamma.a + a.\gamma = 0,$$

$$\gamma^*a + a\gamma^* = 0, \gamma^*a^* + a^*\gamma^* = 0$$

Define now the Fermionic Weyl operator

$$D(\gamma) = exp(\gamma.a^* - a\gamma^*)$$

Clearly, $D(\gamma)$ is a unitary operator since it is the exponential of a skew Hermitian operator. Now,

$$D(\gamma) = 1 + \gamma a^* - a\gamma^* + (1/2)(\gamma a^* - a\gamma^*)^2$$

$$= 1 + \gamma a^* - a\gamma^* - (1/2)(\gamma a^* a\gamma^* + a\gamma^*\gamma a^*)$$

$$= 1 + \gamma a^* - a\gamma^* + (1/2)(\gamma^*\gamma a^*a - \gamma^*\gamma(1 - a^*a))$$

$$= 1 + \gamma a^* - a\gamma^* + \gamma^*\gamma(a^*a - 1/2)$$

Then,

$$aD(\gamma) = a - \gamma aa^* + \gamma^*\gamma a/2 = (1 + \gamma^*\gamma/2)a - \gamma aa^*$$

$$D(\gamma)a = a + \gamma a^*a - \gamma^*\gamma a/2 = (1 - \gamma^*\gamma/2)a + \gamma a^*a$$

Thus,

$$aD(\gamma) - D(\gamma)a = \gamma^*\gamma a - \gamma$$

However,

$$D(\gamma)\gamma = \gamma - \gamma^*\gamma a$$

$$\gamma.D(\gamma) = \gamma - \gamma^*\gamma a$$

ie,

$$[\gamma, D(\gamma)] = 0$$

Thus,

$$aD(\gamma) - D(\gamma)a = -D(\gamma)\gamma = -\gamma.D(\gamma)$$

These equations can be rearranged as

$$D(\gamma)aD(\gamma)^{-1} = a + \gamma,$$

$$D(\gamma)^{-1}.aD(\gamma) = a - \gamma$$

We define the Fermionic single particle coherent state as

$$|\gamma >= D(-\gamma)|0 >= D(\gamma)^{-1}|0 >$$

where $|0>$, is the vacuum, ie, zero particle state. Then

$$a|\gamma >= aD(\gamma)|0 >= D(\gamma)^{-1}.D(\gamma)a.D(\gamma)^{-1}|0 >= D(\gamma)^{-1}(a+\gamma)|0 >$$

$$= \gamma.D(\gamma)^{-1}|0 >= \gamma|\gamma >$$

This proves the desired property of a coherent state, namely that it should be an eigenvector of the annihilation operator. We observe that

$$D(\gamma)^{-1} = 1 - \gamma a^* + a\gamma^* + \gamma^*\gamma(a^*a - 1/2)$$

and hence

$$|\gamma >= |0 > -\gamma|1 > -(1/2)\gamma^*\gamma|0 >= (1 - \gamma^*\gamma/2)|0 > -\gamma|1 > ---(1)$$

From this expression, we can directly verify the coherent state property:

$$a|\gamma >= \gamma.a|1 >= \gamma|0 >,$$

while

$$\gamma|\gamma >= \gamma|0 >$$

since

$$\gamma^2 = 0, \gamma\gamma^*\gamma = -\gamma^*\gamma^2 = 0$$

proving thereby the coherent state property for the state (1).

Now we are in a position to discuss physical implications for Fermionic coherent states. The first observation is that a coherent state is not parametrized by a complex number, it is parametrized by a Fermionic/Grassmannian parameter or a set of anticommuting Grassmannian parameters. Then, if we compute average values of quantities like for example the Dirac four current density in such a state, we will get a Grassmannian number. What physical significance does this have when our averages are not real or complex numbers ? The answer to this question is provided by the Berezin integral: Let $\phi(\gamma,\gamma^*)$ be a function of the Grassmannian parameters γ,γ^* so that the Berezin integral

$$\rho=\int\phi(\gamma,\gamma^*)|\gamma><\gamma|d\gamma.d\gamma^*$$

defines a mixed state. Then, the average value of a function $F(a,a^*)$ of the Fermionic operators a,a^* in the state ρ becomes a complex number to which we can attach physical meaning:

$$Tr(\rho.F(a,a^*))=\int\phi(\gamma,\gamma^*)<\gamma|F(a,a^*)|\gamma>d\gamma.d\gamma^*$$

Another example involving computing average values of the electromagnetic field emitted by a field of electrons and positrons in a given coherent state of the electron-positron field. Let $a_k, k=1,2,...$ denote the annihilation operators of the electrons and positrons after discretizing in momentum space. They satisfy the CAR

$$[a_k,a_m]_+=0,[a_k,a_m^*]=\delta_{km}$$

The current density field generated by this field is according to Dirac's theory, a quadratic function of these operators and hence the electromagnetic field generated by this current density according to the retarded potential formula, is also a quadratic function of these operators. We can express this electromagnetic field as

$$F_{\mu\nu}(x)=\sum_{k,m=1}^{N}[G_{\mu\nu}(x,k,m,1)a_ka_m+\bar{G}_{\mu\nu}(x,k,m,1)a_m^*a_k^*+G_{\mu\nu}(x,k,m,2)a_k^*a_m],x\in\mathbb{R}^4$$

This should be a Hermitian operator field and hence

$$\bar{G}_{\mu\nu}(x,k,m,2)=G_{\mu\nu}(x,m,k,2)$$

The coherent state of the electrons and positrons is given by

$$|\gamma>=D(\gamma)|0>,D(\gamma)=\Pi_{k=1}^{N}exp(\gamma(k)a_k^*-a_k\gamma(k)^*)$$

where $\gamma=((\gamma(k)))_{k=1}^{N}$ are Fermionic/Grassmannian parameters and $\gamma(k)$ and $\gamma(k)^*$ anticommute with $\gamma(l),\gamma(l)^*,a_l,a_l^*$ for all l. We can write

$$D(\gamma)=\Pi_{k=1}^{N}(1+\gamma(k)a_k^*-a_k\gamma(k)^*+\gamma(k)^*\gamma(k)(a_k^*a_k-1/2))$$

The state of the electrons and positrons is assumed to be given by a Berezin integral based superposition of the coherent states:

$$\rho(\theta)=\int\phi(\gamma,\gamma^*|\theta)|\gamma><\gamma|d\gamma.d\gamma^*$$

and hence the average electromagnetic field in this state is

$$< F_{\mu\nu}(x) > (\theta) = Tr(\rho(\theta)F_{\mu\nu}(x)) =$$

$$\int \phi(\gamma, \gamma^*|\theta) < \gamma|F_{\mu\nu}(x)|\gamma > d\gamma.d\gamma^*$$

where

$$< \gamma|F_{\mu\nu}(x)|\gamma >=$$

$$\sum_{k,m=1}^{N} [G_{\mu\nu}(x,k,m,1) < \gamma|a_k a_m|\gamma > +\bar{G}_{\mu\nu}(x,k,m,1) < \gamma|a_m^* a_k^*|\gamma >$$
$$+G_{\mu\nu}(x,k,m,2) < \gamma|a_k^* a_m|\gamma >], x \in \mathbb{R}^4$$

$$= \sum_{k,m=1}^{N} [G_{\mu\nu}(x,k,m,1)\gamma(k)\gamma(m)+\bar{G}_{\mu\nu}(x,k,m,1)\gamma(m)^*\gamma(k)^*$$
$$+G_{\mu\nu}(x,k,m,2)\gamma(k)^*\gamma(m)], x \in \mathbb{R}^4$$

We can now control the parameter vector θ so that this average electromagnetic field is as close as possible to a desired electromagnetic field $F_{d\mu\nu}(x)$ over a given space-time region $x \in D$ by minimizing

$$E(\theta) = \int_D | < F_{\mu\nu}(x) > (\theta) - F_{\mu\nu}(x)|^2 d\mu(x)$$

where $\mu(.)$ is a measure on D.

Remark 1: More generally, we can compute all the statistical moments of the radiation field

$$Tr(\rho(\theta)F_{\mu_1\nu_1}(x_1)...F_{\mu_k\ni_k}(x_k) >$$

in the superposed coherent state $\rho(\theta)$. This computation will involve determining coherent state expectations such as

$$< \gamma|a_{k_1}^*...a_{k_r}^* a_{s_1}...a_{s_m}|\gamma >$$

and noting that this evaluates to

$$\gamma(k_r)^*...\gamma(k_1)^*\gamma(s_1)...\gamma(s_m)$$

The reference for Fermionic coherent state for us has been the master's thesis by Greplova, title "Fermionic Gaussian States".

3.6 Calculating the moments of the radiation field produced by electrons and positrons in the far field when the Fermions are in a coherent state

Remark 2: From Steven Weinberg's book, "The quantum theory of fields, vol.1", it is known that the free Dirac field can be expanded in terms of momentum-spin space electron annihilation operators $a(P,\sigma)$ and positron creation operators $b(P,\sigma)^*$ which satisfy the CAR (canonical anticommutation relations)

$$[a(P,\sigma), a(P',\sigma')^*]_+ = \delta^3(P-P')\delta_{\sigma,\sigma'},$$

$$[b(P,\sigma), b(P',\sigma')^*]_+ = \delta^3(P-P')\delta_{\sigma,\sigma'}$$

and all the other anticommutators evaluating to zero. The second quantized Dirac wave field is then the solution to Dirac's relativistic wave equation and is given by

$$\psi(x) = \psi(t,r) = \int [a(P,\sigma)u(P,\sigma)exp(-ip.x) + b(P,\sigma)^* v(P,\sigma)exp(ip.x)]d^3P$$

where

$$p^0 = E(P) = \sqrt{m^2 + P^2}$$

The Dirac current density operator field is then

$$J^\mu(x) = -e\psi(x)^*\gamma^0\gamma^\mu\psi(x)$$

and it is evident that this can be expressed as a linear combination of the quadratic operators

$$a(P,\sigma)^*a(P,\sigma'), a(P,\sigma)^*b(P,\sigma')^*, b(P,\sigma)a(P,\sigma')^*, b(P,\sigma)b(P',\sigma')^*$$

Thus, using the retarded potential formula for the Maxwell equations in the form

$$A^\mu(x) = \int G(x-x')J^\mu(x')d^4x'$$

it is evident that once again $A^\mu(x)$ is expressible as a linear combination of the above quadratic operators. After discretizing the integrals in 3-momentum space, we then club all the electron and positron annihilation operators into one set $\{a_k\}$ and their adjoints into $\{a_k^*\}$ and then use the above coherent state formalism of Greplova to determine the quantum averages of the electromagnetic field.

3.7 Controlling the classical em fields interacting with the Dirac field so that the mean value of the em field radiated by the resulting Dirac second quantized current in a Fermionic coherent state is as close as possible to a given deterministic pattern in space and simultaneously the mean square fluctuations of this field in a Fermionic coherent state are minimized

The ultimate aim of all these computations can be formulated in very simple terms as an optimization problem: Design the control parameters θ or the control classical fields to a quantum antenna so that the error energy between the average value of the quantum electromagnetic field produced by the quantum antenna and the desired classical electromagnetic field pattern is a minimum subject to the constraint that the second order central moments of the quantum electromagnetic field (ie variance of fluctuations) is smaller than a given threshold.

Remark:More generally, we can control the wave function operator of the Dirac field of electrons and positrons as well as the Maxwell photon field operators within the cavity resonator antenna by introducing classical control current and electromagnetic field sources into the cavity. The quantum cavity photon and electron-positron fields will then be expressible in terms of the free quantum fields plus additional perturbation terms involving the classical current and field sources. Once this is done, we can in principle calculate the far field antenna pattern produced by the cavity surface currents induced by the tangential components of the quantum magnetic field operators as well as that produced by the Dirac field of electrons and positrons and then design these classical control fields so that the far field quantum Poynting radiation pattern has a mean value and correlations in a given quantum coherent state of the photons and electrons-positrons within the cavity as close as possible to specified values.

To formulate this optimization problem in abstract terms, let

$$X(t,r) = X(x) = \sum_k [f_k(\theta)X_k(t,r) - \bar{f}_k(\theta)X_k(t,r)^*]$$

be the quantum field radiated out by the quantum antenna where θ is a control parameter vector or a classical field. This form arises typically by perturbatively solving the Maxwell-Dirac field equations upto linear orders in the perturbing classical current and electromagnetic field. $f_k(\theta)$ is a complex valued function of θ and $X_k(t,r)$ are quantum operator fields. The average value of this radiated field in a state $|\Phi>$ is given by

$$< X(t,r) >=< \Phi|X(t,r)|\Phi >=$$

$$\sum_k [f_k(\theta) < \Phi|X_k(t,r)|\Phi > +\bar{f}_k(\theta) < \Phi|X_k(t,r)^*|\Phi >]$$

and the central correlations in the field are

$$C(t,r|t',r') =< X(t,r).X(t',r')^* > - < X(t,r) >< X(t',r')^* >$$

$$= \sum_{k,m} f_k(\theta)\bar{f}_m(\theta) < \Phi|(X_k(t,r)- < X_k(t,r) >).(X_m(t',r')^*- < X_m(t',r') >^* |\Phi >$$

$$+ \sum_{k,m} f_k(\theta)f_m(\theta) < \Phi|(X_k(t,r)- < X_k(t,r) >).(X_m(t',r')- < X_m(t',r') >)|\Phi >$$

$$+c.c$$

where $c.c$ denotes complex conjugate of the previous terms. It is then easy to see that if the $f_k's$ are linear functions of θ, then the problem of minimizing $\int_D (< X(t,r) > -X_d(t,r))^2 dtd^3r$ subject to the constraint that $\int_{D\times D} W(t,r|t',r')C(t,r|t',r')dtd^3rdt'd^3r'$ is fixed is equivalent to finding the minimum of the ratio of two quadratic forms:

$$E(\theta) = \frac{\theta^T Q_1 \theta}{\theta^T Q_2 \theta}$$

where Q_1, Q_2 are two Hermitian positive definite matrices and the optimal equations for θ then result in the generalized eigenvalue problem

$$(Q_1 - cQ_2)\theta = 0$$

The minimum value of this ratio is the minimum of all the generalized eigenvalues c, ie, the minimum of all the $c's$ for which

$$det(Q_1 - cQ_2) = 0$$

and the the optimal value of θ is a generalized eigenvector corresponding to the minimum of c with the normalization condition

$$\theta^T Q_2 \theta = E$$

where E is the prescribed value of the energy of the quantum fluctuations $\int_{D\times D} W(t,r|t',r')C(t,r|t',r')dtd^3rdt'd^3r'$.

3.8 Approximate analysis of a rectangular quantum antenna

The quantum antenna is assumed to be the cuboid region $[0, a] \times [0, b] \times [0, d]$. This rectangular cavity is assumed to comprise of photons, electrons and positrons. The exact equations governing the quantum fields corresponding to these particles are (a) The Maxwell equations for the four vector potential driven by the Dirac field current and (b) The Dirac field equations driven by an interaction between the Dirac field and the Maxwell field four vector potential. These exact field equations are:

$$\Box A_\mu(x) = \mu_0 e\psi^*(x)\alpha^\mu\psi(x), x = (t, r) - -(1)$$

$$((\alpha, -i\nabla) + \beta m_0)\psi(t, r) + eA_\mu(x)\alpha^\mu\psi(x) = i\partial_t\psi(t, r) - - - (2)$$

where

$$\alpha^\mu = \gamma^0\gamma^\mu, \beta = \gamma^0$$

and γ^μ are the Dirac Gamma matrices. Note that $(\gamma^0)^2 = I_4$ and hence $\alpha^0 = I_4$. The boundary conditions under which we need to solve these Maxwell-Dirac equations are that the Dirac operator wave field $\psi(x)$, the tangential components of the electric field $F_{0r} = A_{r,0} - A_{0,r}, r = 1, 2, 3$ and the normal components of the magnetic field $F_{rs} = A_{s,r} - A_{r,s}, 1 \leq r < s \leq 3$ must vanish on the boundaries of the cavity. In particular, the freed Dirac field must have an expansion

$$\psi^{(0)}(t, r) = \sum_{mnp} \mathbf{c}(mnp, t)u_{mnp}(r)$$

where m, n, p run over positive integers and

$$u_{mnp}(r) = (2\sqrt{2}/\sqrt{abd})sin(m\pi x/a)sin(n\pi y/b)sin(p\pi z/d)$$

Substituting this into the free Dirac equation, ie, without any electromagnetic interactions, we get

$$\sum_{mnp}(i\partial_t c(mnp, t))u_{mnp}(r) =$$

$$((\alpha, -i\nabla) + \beta m_0). \sum_{mnp} \mathbf{c}(mnp, t)u_{mnp}(r)$$

from which we derive on taking the inner products on both sides with $u_{kls}(r)$ and using the orthonormality of this set of functions over the cavity volume, ie,

$$< u_{kls}, u_{mnp} >= \int_B u_{kls}(r)u_{mnp}(r)d^3r = \delta_{km}\delta_{ln}\delta_{sp}$$

where B is the cavity volume

$$B = [0, a] \times [0, b] \times [0, d],$$

the following sequence of differential equations

$$i\partial_t c(kls,t) = \sum_{mnp} [< u_{kls}, -i\partial_x u_{mnp} > \alpha_1 c(mnp,t)$$

$$+ < u_{kls}, -i\partial_y u_{mnp} > \alpha_2 c(mnp,t) + < u_{kls}, -i\partial_z u_{mnp} > \alpha_3 c(mnp,t)] + m_0\beta c(kls,t)$$

Now we evaluate

$$< u_{kls}, -i\partial_x u_{mnp} >= -i\delta_{ln}\delta_{sp}(m\pi/a)(\int_0^a (2/a)sin(k\pi x/a)cos(m\pi x/a)dx$$

$$= a_1(k,m)\delta_{ln}\delta_{sp}$$

where

$$a_1(k,m) = (-2im\pi/a^2)\int_0^a sin(k\pi x/a).cos(m\pi x/a)dx$$

Likewise,

$$< u_{kls}, -i\partial_y u_{mnp} >= a_2(l,n)\delta_{km}\delta_{sp},$$

and

$$< u_{kls}, -i\partial_z u_{mnp} >= a_3(s,p)\delta_{km}\delta_{ln}$$

Combining all these equations gives us finally,

$$i\partial_t c(kls,t) = \sum_m a_1(k,m)\alpha_1 c(mls,t)$$

$$+\sum_n a_2(l,n)\alpha_2 c(kns,t) + \sum_p a_3(s,p)\alpha_3 c(klp,t)] + m_0\beta c(kls,t)$$

Note that $\alpha_1, \alpha_2, \alpha_3, \beta$ are 4×4 Hermitian matrices while $c(mnp,t)$ is a 4×1 complex vector. Arranging the 4×1 vectors $c(mnp,t), m,n,p \geq 1$ in lexicographic order to give an infinite vector $\mathbf{c}(t)$ and likewise defining a block structured infinite dimensional Dirac Hamiltonian matrix $\mathbf{H}_0$ by

$$\mathbf{H}_0 = \sum_{klsm} a_1(k,m)(\mathbf{I}_4 \otimes \mathbf{e}(kls))\alpha_1(\mathbf{I}_4 \otimes \mathbf{e}(mls)^T)$$

$$+\sum_{lksn} a_2(l,n)(\mathbf{I}_4 \otimes \mathbf{e}(kls))\alpha_2(\mathbf{I}_4 \otimes \mathbf{e}(kns)^T)$$

$$+\sum_{klsp} a_3(s,p)(\mathbf{I}_4 \otimes \mathbf{e}(kls))\alpha_3(\mathbf{I}_4 \otimes \mathbf{e}(klp)^T)$$

$$+m_0\beta \otimes \mathbf{I}$$

where we may choose $\mathbf{e}(mnp), m,n,p \geq 1$ as any orthonormal basis for $l^2(\mathbb{Z}_+)$, the Hilbert space of all one sided square summable infinite sequences and define

$$\mathbf{c}(t) = \sum_{mnp} c(mnp,t)\mathbf{e}(mnp)$$

By orthonormal, we mean that

$$\mathbf{e}(kls)^T\mathbf{e}(mnp) = \delta_{km}\delta_{ln}\delta_{sp}$$

Thus the free Dirac equation in the RDRA has been put in "Standard" block matrix form:

$$i\frac{d\mathbf{c}(t)}{dt} = \mathbf{H}_0\mathbf{c}(t)$$

the general solution to which can be expressed as

$$\mathbf{c}(t) = \sum_n d(n).\mathbf{c}_n exp(-iE(n)t)$$

where $\mathbf{c}_n, n \geq 1$ form an orthonormal basis for $l^2(\mathbb{Z}_+)$ and the $d(n)'s$ are arbitrary complex numbers such that

$$\sum_n |d(n)|^2 = 1$$

$E(n)'s$ are the (energy) eigenvalues of the infinite dimensional Hermitian H_0:

$$det(\mathbf{H}_0 - E(n)\mathbf{I}) = 0$$

The average energy of the free Dirac field of electrons and positrons within the cavity is then

$$< \mathbf{c}(t), \mathbf{H}_0\mathbf{c}(t) >= \sum_n E(n)d(n)^*d(n)$$

It is easy to see as in the case of the Dirac equation in free space that if $E(n)$ is an eigenvalue of $\mathbf{H}_0$ then so is $-E(n)$ where the $E(n)'s$ may be taken as positive, Hence if $\mathbf{c}_{en}$ is an eigenvector of $\mathbf{H}_0$ corresponding to the eigenvalue $E(n)$ and $\mathbf{c}_{pn}$ is an eigenvector corresponding to the eigenvalue $-E(n)$, then the solution can be expressed as

$$\mathbf{c}(t) = \sum_n [d_e(n)\mathbf{c}_{en}exp(-iE(n)t) + d_p(n)^*\mathbf{c}_{pn}exp(iE(n)t)]$$

Therefore, it is plausible in the second quantized theory, to look upon the $d_e(n)'s$ as annihilation operators of the electrons and the $d_p(n)^*$'s as the creation operators of the positrons. The actual Dirac wave function $\psi(t,r)$ in the absence of electromagnetic interactions is then

$$\psi(t,r) = \sum_{kmnp} [d_e(k)c_{ek}(mnp)u_{mnp}(r)exp(-iE(k)t)+$$

$$d_p(k)^*c_{pk}(mnp)u_{mnp}(r)exp(iE(k)t)] --- (3)$$

A simple calculation then shows that the second quantized Hamiltonian of the free Dirac field of electrons and positrons within the cavity is given by

$$H_{D0} = \int_B \psi(t,r)^*((\alpha, -i\nabla) + \beta m)\psi(t,r)d^3r$$

$$= \int_B \psi(t,r)^* i\partial_t \psi(t,r) d^3r$$

$$= \sum_k E(k)(d_e(k)^* d_e(k) - d_p(k) d_p(k)^*)$$

Now from the basic anticommuation relations for the Dirac field, we have

$$\{\psi(t,r), \psi(t,r')^*\} = \delta^3(r - r')I$$

and this immediately implies the following anticommutation relations for the electron and positron creation and annihilation operators:

$$\{d_e(k), d_e(m)^*\} = \delta_{km}, \{d_p(k), d_p(m)^*\} = \delta_{km}$$

with all the other anticommutators vanishing. This completes our description of the free Dirac field of electrons and positrons within the RDRA. Using these anticommutation relations, we immediately get that the total second quantized Hamiltonian of the free Dirac field in the cavity can equivalently be expressed as

$$H_{D0} = \sum_k E(k)(d_e(k)^* d_e(k) + d_p(k)^* d_p(k))$$

namely, the sum of the total electron and positron energies. Likewise, when we solve the free Maxwell equations within the cavity after incorporating the appropriate boundary conditions, we get that the scalar potential is zero since there are no charges while the magnetic vector potential admits an expansion obtained from

$$\mathbf{E} = -\partial_t \mathbf{A}$$

as

$$\mathbf{A}(t,r) = \sum_k [b(k)\mathbf{w}_k(r) exp(-i\omega(k)t) + b(k)^* \mathbf{w}_k(r)^* exp(i\omega(k)t)] \quad --- (4)$$

where now $\mathbf{w}_k(r)$ has three components that are calculated from the expansion of E_z and the relationship between the transverse and longitudinal components of the electric field within the cavity. Note that the electromagnetic field is being computed in the Coulomb gauge which implies that the electric scalar potential becomes zero in view of the fact that in the Coulomb gauge, the scalar potential satisfies Poisson's equation and is therefore a matter field which evaluates to zero since there is no unperturbed charge density.

We also note that the third, ie, z component of $\mathbf{w}_k(r)$ where the index k is identified with the modal triplet (mnp) is proportional to $sin(m\pi x/a) sin(n\pi y/b) cos(p\pi z/d)$ in view of the boundary conditions on the electric field an the fact that each mode of the magnetic vector potential is proportional to the electric field $(-j\omega\mathbf{A} = \mathbf{E})$. $b(k) = b(mnp)$ is identified with a photon annihilation operator while $b(k)^*$ with a photon creation operator. They satisfy the canonical commutation relations

$$[b(k), b(m)^*] = \delta_{km}$$

Formally, we can compute both the free Dirac current density $\psi(t,r)^*\alpha^\mu\psi(t,r)$ of electrons and positrons within the cavity as well as the surface current density on the RDRA walls induced by the tangential components of the quantum magnetic field $\mathbf{B} = curl\mathbf{A}$ and obtain the far field radiation pattern generated by both of these cavity current components. Obviously, this far field radiation pattern will have its first component being a quadratic form in the electron-positron creation and annihilation operators $d_e(k), d_p(k), d_e(k)^*, d_p(k)^*$ while the second component will be linear in the photon creation-annihilation operators $b(k), b(k)^*$ and therefore, in principle, we can compute all the statistical moments of the radiation field in a joint coherent state of the photons, electrons and positrons. However, this picture of the far field quantum radiation pattern is incomplete because it does not take into account the cavity current density terms caused by perturbation in the Dirac wave field due to interaction with the photons and it does not also take into account the cavity surface current density terms caused by perturbation in the Maxwell field caused by its interaction with the Dirac field. We shall now indicate an approximate first order calculation by which these extra correction terms may be obtained due interactions between the Maxwell field and the Dirac field.

We denote the free Dirac field within the cavity derived above by $\psi^{(0)}(t,r)$ and the corresponding momentum space wave function $c(mnp,t)$ by $c^{(0)}(mnp,t)$. Likewise, we denote the free Maxwell field within the cavity by $\mathbf{A}^{(0)}$. Let $\delta\mathbf{A}$ denote the perturbation to the Maxwell field caused by the Dirac current and $\delta\psi, \delta c(mnp,t)$ the perturbation to the Dirac field caused by the Maxwell current. Then, clearly if $S(x-y)$ denotes the electron propagator and $D(x-y)$ the photon propagator, we have using (1) and (2), approximately,

$$\delta A^\mu(t,r) = \mu_0 e \int D(t-t', r-r')\psi^{(0)*}(t',r')\alpha^\mu\psi^{(0)}(t',r')dt'd^3r'$$

$$\delta\psi(t,r) = e\int S(t-t',r-r')A_\mu^{(0)}(t',r')\alpha^\mu\psi^{(0)}(t',r')dt'd^3r'$$

where we substitute for $\psi^{(0)}$ and $A_\mu^{(0)}$ the expressions given in (3) and (4). Then,

$$\psi^{(0)*}\alpha^\mu\psi^{(0)}(t,r) =$$

$$\sum_{kmnpk'm'n'p'} [d_e(k)^*\bar{c}_{ek}(mnp)exp(iE(k)t) + d_p(k)\bar{c}_{pk}(mnp)exp(-iE(k)t)].\alpha^\mu.$$

$$.[d_e(k')c_{ek'}(m'n'p')exp(-iE(k')t)+d_p(k')^*c_{pk'}(m'n'p')exp(iE(k')t)]u_{mnp}(r)u_{m'n'p'}(r)$$

$$= \sum[d_e(k)^*d_e(k')\bar{c}_{ek}(mnp)\alpha^\mu c_{ek'}(m'n'p')exp(i(E(k)-E(k'))t)u_{mnp}(r)u_{m'n'p'}(r)]$$

$$+\sum[d_e(k)^*d_p(k')^*\bar{c}_{ek}(mnp)\alpha^\mu c_{pk'}(m'n'p')exp(i(E(k)+E(k'))t)u_{mnp}(r)u_{m'n'p'}(r)]$$

$$+\sum[d_p(k)d_e(k')\bar{c}_{pk}(mnp)\alpha^\mu c_{ek'}(m'n'p')exp(-i(E(k)+E(k'))t)u_{mnp}(r)u_{m'n'p'}(r)]$$

$$+\sum[d_p(k)d_p(k')^*\bar{c}_{pk}(mnp)\alpha^\mu c_{ek'}(m'n'p')exp(-i(E(k)-E(k'))t)u_{mnp}(r)u_{m'n'p'}(r)]$$

We see that the frequencies of the Dirac current that generate the perturbation to the quantum electromagnetic field are $E(k) \pm E(k'), k, k' = 1, 2, ...$ or more precisely, these divided by Planck's constant. Here $E(k)$ was obtained by solving the free Dirac eigenvalue equation inside the rectangular cavity with zero boundary conditions. The $E(k)'s$ were obtained as the eigenvalues of the Dirac Hamiltonian. From basic principles of special relativity, it is easy to see that these $E(k)'s$ are of the order

$$c\sqrt{m_0^2c^2 + P^2}$$

where

$$P^2 = (h/2\pi)^2((m\pi/a)^2 + (n\pi/b)^2 + (p\pi/d)^2)$$

with m, n, p being positive integers determined by the mode of oscillation of the field within the cavity. Now this current is of the general form

$$-e\psi^{(0)*}\alpha^\mu\psi^{(0)}(t,r)$$

$$= \sum_{k,k'}[d(k)^*d(k')f^\mu_{kk'}(t,r) + d(k)d(k')g^\mu_{kk'}(t,r) + d(k)^*d(k')^*\bar{g}^\mu_{kk'}(t,r)]$$

where the $d(k)'s$ are the annihilation operators of the electrons and positrons and their adjoints $d(k)^*$ are the corresponding creation operators. The functions $f_{kk'}, g_{kk'}$ are constructed by superposing $exp(\pm i(E(k)\pm E(k'))u_{mnp}(r)u_{m'n'p'}(r)$ and these components are easily seen to be expressible as superpositions of space-time sinusoids with the temporal frequencies being $E(k) \pm E(k')$ or their negatives and the spatial frequencies, ie, wave-numbers being $m\pi/a, n\pi/b, p\pi/d, m'\pi/a, n'\pi/b, p'\pi/d$. Note that because the current density is a Hermitian operator field, it follows that

$$\bar{f}^\mu_{kk'}(t,r) = f^\mu_{k'k}(t,r)$$

Then, the perturbation in the electromagnetic potentials can be expressed as

$$\delta A^\mu(t,r) =$$

$$\sum_{k,k'}[d(k)^*d(k')\int D(t-t',r-r')f^\mu_{kk'}(t',r')dt'd^3r' + d(k)d(k')\int D(t-t',r-r')g^\mu_{kk'}(t',r')dt'd^3r'$$

$$+d(k)^*d(k')^*\int D(t-t',r-r')\bar{g}^\mu_{kk'}(t',r')dt'd^3r']$$

Let

$$J^\mu(k) = -e\int \psi^{(0)*}(x)\alpha^\mu\psi^{(0)}(x)exp(-ik.x)d^4x$$

$$= \int \psi^{(0)*}(t,r)\alpha^\mu\psi^{(0)}(t,r)exp(-i(k^0t - K.r))dtd^3r$$

where

$$k = (k^\mu) = (k^0, K)$$

denote the space-time four dimensional Fourier transform of the unperturbed Dirac four current density. Then, we can write down the space-time Fourier transform of the correction $\delta A^\mu(x), x = (t,r)$ to the electromagnetic four potential caused by this Dirac current as

$$\delta A^\mu(k) = \int \delta A^\mu(x).exp(-ik.x)d^4x$$

$$= \mu_0 D(k)J^\mu(k) = \mu_0 J^\mu(k)/k^2, k^2 = k_\mu k^\mu = (k^0)^2 - |K|^2$$

in units where $c = 1$. It should be noted that by the convolution theorem for Fourier transforms, if $\psi^{(0)}(k)$ denotes the space-time Fourier transform of $\psi^{(0)}(x)$, then

$$J^\mu(k) = (2\pi)^{-4}\int \psi^{(0)*}(k'-k)\alpha^\mu\psi^{(0)}(k')d^4k'$$

and hence, the perturbation to the electromagnetic four potential in the space-time Fourier domain, ie, in four momentum space of the photon can be expressed as

$$\delta A^\mu(k) = (\mu_0/(2\pi)^4k^2)\int \psi^{(0)*}(k'-k)\alpha^\mu\psi^{(0)}(k')d^4k'$$

3.9 Remark on the perturbation in the quantum Dirac field and the quantum electromangetic field interacting with each other caused by further interaction of the Dirac field with a classical control em field and interaction of the quantum electromagnetic field with a control classical current

The unperturbed electromagnetic field is in the Coulomb gauge, ie, $div\mathbf{A}^{(0)} = 0$ and also since there is no charge/current for the unperturbed field, the unperturbed electric scalar potential is a matter field which is identically zero, ie, $A^{(0)0} = 0$. Hence, we are guaranteed that the unperturbed electromagnetic potentials also satisfy the Lorentz gauge conditions, ie, $div\mathbf{A}^{(0)} + \partial_t A^{(0)0} = 0$. This means that while computing the perturbations to the electromagnetic potentials caused by currents coming from the Dirac field, we can safely work in the Lorentz gauge.

Likewise, the change in the Dirac field caused by interaction with the electromagnetic field within the cavity is given upto first order perturbation theory by

$$\delta\psi(x) = \delta\psi(x) = e\int S(x-x')A_\mu^{(0)}(x')\alpha^\mu\psi^{(0)}(x')d^4x'$$

$$= e\int S(x-x')A_r^{(0)}(x')\alpha^r\psi^{(0)}(x')d^4x'$$

$$= -e\int S(x-x')(\alpha, \mathbf{A}^{(0)}(x'))\psi^{(0)}(x')d^4x'$$

$$= -e\int S(t-t', r-r')\sum_k [b(k)(\alpha, \mathbf{w}_k(r'))exp(-i\omega(k)t')+b(k)^*(\alpha, \mathbf{w}_k(r')^*)exp(i\omega(k)t')].$$

$$.[\sum_{kmnp}[d_e(k)c_{ek}(mnp)u_{mnp}(r')exp(-iE(k)t')+d_p(k)^*c_{pk}(mnp)u_{mnp}(r')exp(iE(k)t')]dt'd^3r'$$

$$= -e\sum_{kk'mnp} b(k)d_e(k')\int S(t-t', r-r')(\alpha, \mathbf{w}_k(r'))c_{ek'}(mnp)u_{mnp}(r')exp(-i(\omega(k)+E(k'))t')dt'd^3r'$$

$$-e\sum_{kk'mnp} b(k)d_p(k')^*\int S(t-t', r-r')(\alpha, \mathbf{w}_k(r')^*)c_{pk'}(mnp)u_{mnp}(r')exp(-i(\omega(k)-E(k'))t')dt'd^3r'$$

$$-e\sum_{kk'mnp} b(k)^*d_e(k')\int S(t-t', r-r')(\alpha, \mathbf{w}_k(r'))c_{ek'}(mnp)u_{mnp}(r')exp(i(\omega(k)-E(k'))t')dt'd^3r'$$

$$-e\sum_{kk'mnp} b(k)^*d_p(k')^*\int S(t-t', r-r')(\alpha, \mathbf{w}_k(r')^*)c_{pk'}(mnp)u_{mnp}(r')exp(i(\omega(k)+E(k'))t')dt'd^3r'$$

From this expression, it is clear that the characteristic frequencies of the interaction term between the electromagnetic potentials and the Dirac field and hence the characteristic frequencies of the perturbation in the Dirac field caused by electromagnetic interaction are $\pm\omega(k) \pm E(k')$. In terms of the compact notation introduced above, namely using the same symbol $d(k)$ for both electron and positron annihilation operators and likewise $d(k)^*$ for both electron and positron creation operators, we can write

$$\delta\psi(x) = \int S(t-t', r-r')[\sum b(k)d(k')h_{1kk'}(t', r') + b(k)d(k')^*h_{2kk'}(t', r') +$$

$$+ b(k)^*d(k')h_{3kk'}(t', r') + b(k)^*d(k')^*h_{4kk'}(t', r')]dt'd^3r'$$

where the functions $h_{mkk'}(t, r)$ are built by superposing the functions $exp(i \pm (\omega(k) \pm E(k'))t)(\alpha, w_k(r))c_{k'}(mnp)u_{mnp}(r)$ and the same expression with $w_k(r)$ replaced by its complex conjugate $w_k(r)^*$. Here, the symbol $c_{k'}(mnp)$ stands for either $c_{ek'}(mnp)$ or $c_{pk'}(mnp)$.

In particular this expression shows that the perturbation to the Dirac field caused by electromagnetic interactions have frequencies $\pm\omega(k) \pm E(k')$, namely linear combinations of the unperturbed electromagnetic characteristic frequencies and the unperturbed Dirac characteristic frequencies. This represents a new feature of our model.

Before proceeding further, observe that we can write in the four dimensional momentum/space-time frequency domain,

$$\delta\psi(k) = S(k)\mathcal{F}(eA_\mu^{(0)}\alpha^\mu\psi^{(0)})(k)$$

where

$$S(k) = (k^0 - (\alpha, \mathbf{K}) - \beta m_0 + i0)^{-1}$$

is the electron propagator in the four momentum domain $k = (k^\mu) = (k^0, \mathbf{K})$ and

Control of the quantum electromagnetic field and the Dirac field of electrons and positrons within the rectangular cavity by means of a classical electromangetic field coming from a laser source connected to the cavity plus a classical current source coming from a probe inserted into the cavity: Let $A_\mu^c(x)$ denote the classical electromagnetic four potential from the laser and $J_\mu^c(x)$ the classical current density coming from the probe insertion. The relevant equations are

$$\Box\mathbf{A}_\mu = -e\mu_0\psi^*\alpha_\mu\psi + \mu_0\mathbf{J}_\mu^c,$$

$$((\alpha, -i\nabla) + \beta m)\psi = [-e(\alpha, \mathbf{A}) - e(\alpha, \mathbf{A}^c)]\psi$$

The first order perturbative solution to these equations is with $x = (t, \mathbf{r})$,

$$\psi(x) = \psi^{(0)}(x) + \delta\psi(x),$$

$$A^r(x) = A^{r(0)}(x) + \delta A^r(x), r = 1, 2, 3$$

where

$$\psi^{(0)}(x) =$$

$$\sum_{kmnp}[d_e(k)c_{ek}(mnp)u_{mnp}(r)exp(-iE(k)t) + d_p(k)^*c_{pk}(mnp)u_{mnp}(r)exp(iE(k)t)]$$

$$A^{r(0)}(x) = \sum_k[b(k)w_k^r(\mathbf{r})exp(-i\omega(k)t) + b(k)^*\bar{w}_k^r(\mathbf{r})exp(i\omega(k)t)]$$

$$\delta\psi(x) =$$

$$-e\int S_e(x-y)[(\alpha, \mathbf{A}^{(0)}(y)) + (\alpha, \mathbf{A}^c(y))]\psi^{(0)}(y)d^4y$$

$$= -e\int S_e(x-y)[\alpha^r A_r^{(0)}(y) + \alpha^r A_r^c(y)]\psi^{(0)}(y)d^4y$$

$$= \delta\psi_1(x) + \delta\psi_{ctr}(x),$$

$$\delta A_r(x) = -e\mu_0\int D(x-y)(\psi^*\alpha_r\psi)(y)d^4y + \mu_0\int D(x-y)J_r^c(y)d^4y$$

$$\approx -e\mu_0\int D(x-y)(\psi^{(0)*}\alpha_r\psi^{(0)})(y)d^4y + \mu_0\int D(x-y)J_r^c(y)d^4y$$

where the classically controllable part of the Dirac field is

$$\delta\psi_{ctr}(x) = -e\int S_e(x-y)\alpha^r A_r^c(y)\psi^{(0)}(y)d^4y$$

and this component contains a classical field component A_r^c and a quantum field component $\psi^{(0)}$, while the part of the Dirac field perturbation that is not controllable is

$$\delta\psi_1(x) = -e\int S_e(x-y)\alpha^r A_r^{(0)}(y)\psi^{(0)}(y)d^4y$$

On the other hand, the controllable part of the electromagnetic field is purely classical:

$$\delta A_{r,ctr}(x) = \mu_0\int D(x-y)J_r^c(y)d^4y$$

If we go one step further in the perturbation series, then we get an additional term in the controllable part of the electromagnetic field so that the above equation gets modified to:

$$\delta A_{r,ctr}(x) = \mu_0\int D(x-y)J_r^c(y)d^4y$$

$$-e\mu_0\int D(x-y)(\delta\psi_{ctr}(y)^*\alpha_r(\psi^{(0)}+\delta\psi_1)(y)d^4y$$

$$-e\mu_0\int D(x-y)(\psi^{(0)*}+\delta\psi_1^*)(y)\alpha_r\delta\psi_{ctr}(y)d^4y$$

Note that in this analysis, the perturbation parameter is the electron charge e and if we neglect $O(e^2)$ terms, then the above expression for the controllable part of the electromagnetic field simplifies to

$$\delta A_{r,ctr}(x) = \mu_0\int D(x-y)J_r^c(y)d^4y+$$

$$-e\mu_0\int D(x-y)\delta\psi_{ctr}(y)^*\alpha_r\psi^{(0)}(y)d^4y$$

$$-e\mu_0\int D(x-y)\psi^{(0)*}(y)\alpha_r\delta\psi_{ctr}(y)d^4y$$

In the particular case of the rdra considered here, we find that the controllable part of the Dirac field has the expansion

$$\delta\psi_{ctr}(x) = -e\int S_e(x-y)\alpha^r A_r^c(y)\psi^{(0)}(y)d^4y$$

$$= -e\int S_e(t-t',r-r')\alpha^r A_r^c(t',r')[\sum_{kmnp}[d_e(k)c_{ek}(mnp)u_{mnp}(r')exp(-iE(k)t')+$$

$$d_p(k)^*c_{pk}(mnp)u_{mnp}(r')exp(iE(k)t')]dt'd^3r'$$

Now define the following Fourier components of the control classical laser generated electromagnetic field w.r.t the cavity boundary conditions and the energy spectrum of the free Dirac field in the cavity after:

$$\int A_r^c(t',r')u_{mnp}(r')exp(-iK.r')exp(i\omega t')d^3r'dt' = C_{A,r}(\omega,K|m,n,p)$$

Then, we can express the above controllabe part of the Dirac field in the following form in the spatio-temporal Fourier domain:

$$\int \delta\psi_{ctr}(t,r).exp(i(\omega t - K.r)dtd^3r =$$

$$\sum_{mnpk}[-ed_e(k)S(\omega,K)c_{ek}(mnp)\alpha^r[C_r(\omega - E(k),K|mnp)$$

$$-ed_p(k)^*S(\omega,K)c_{pk}(mnp)\alpha^r C_r(\omega+E(k),K|mnp)]$$

$$= -eS(\omega,K)\sum_{mnpk}[d_e(k)c_{ek}(mnp)\alpha^r C_r(\omega - E(k),$$

$$K|mnp)+d_p(k)^*c_{pk}(mnp)\alpha^r C_r(\omega+E(k)|mnp)]$$

The controllable part of the Dirac four current density is then given upto first order perturbation terms by $(x = (t,r))$

$$\delta J^\mu(t,r) = -e\psi^{(0)*}(x)\alpha^\mu\delta\psi_{ctr}(x) - e\delta\psi_{ctr}(x)^*\alpha^\mu\psi^{(0)}(x)$$

and it is immediately clear from the above expression that the far field radiated electromagnetic potential generated by this controllable current field can be expressed in the form

$$\delta A_R^\mu(t,r) = \int D(t-t',r-r')\delta J^\mu(t',r')dt'd^3r'$$

$$= \sum_{mnpkrm'n'k's} d_e(k)^*d_e(k')\int C_r(\omega - E(k),K|mnp)\bar{C}_s(-\omega' - E(k'),K'|m'n'p')$$

$$.F^{\mu rs}(t,r|\omega,K,\omega',K',mnpk,m'n'p'k')d\omega d^3Kd\omega' d^3K'$$

plus three other similar terms involving $d_e(k)^*d_p(k')^*, d_p(k)d_e(k'), d_p(k)d_p(k')^*$. In compact notation, the expected value of this controllable far field pattern can be expressed as a Hermitian quadratic form in the complex numbers $C_r(\omega,K|mnp), \omega \in \mathbb{R}, K \in \mathbb{R}^3, m,n,p \in \mathbb{Z}_+$. These complex numbers are controllable since they represent in some sense the spatio-temporal components of the Fourier components of the classical control electromagnetic field A_μ^c.

3.10 Quantum Antennas constructed using supersymmetric field theories

Reference: Steven Weinberg, "The quantum theory of fields, vol.III, Supersymmetry", Cambridge University Press.

[1] Let Φ be a left Chiral field, ie, it is a function of only $\theta_L = (1+\gamma_5)\theta/2$ and

$$x_+^\mu = x^\mu + (1/2)\theta_R^T\epsilon\gamma^\mu\theta_L$$

Let V^A be gauge super-fields for each Yang-Mills gauge group index A. Expand V^A as

$$V^A(x,\theta) = \theta^T\epsilon\gamma^\mu\theta.V_\mu^A(x) + \theta^T\epsilon\theta.\theta^T\gamma_5\epsilon\lambda^A(x) + (\theta^T\epsilon\theta)^2 D^A(x)$$

V_μ^A is called the gauge field, λ^A is called the gaugino field and D^A is called the auxiliary field. The transformation law of the gauge superfield under extended gauge transformations defined by an arbitrary left Chiral superfield Ω is given by

$$\Gamma \to exp(i\Omega)\Gamma.exp(-i\Omega^*)$$

Now define a left Chiral spinor supefield

$$W_L = D_R^T\epsilon D_R exp(t.V)D_L(exp(-t.V))$$

where $t.V = t_A V^A$. Note that the gauge superfield transformation law implies

$$exp(t.V) \to exp(i\Omega)exp(t.V).exp(-i\Omega^*)$$

$$exp(-t.V) \to exp(i\Omega^*)exp(-t.V)exp(-i\Omega)$$

Then since Ω is left Chiral, Ω^* becomes right Chiral and therefore under the gauge transformation,

$$W_L \to exp(i\Omega)D_R^T\epsilon D_R exp(t.V)D_L(exp(-t.V)exp(-i\Omega))$$

$$= exp(i\Omega)D_R^T\epsilon D_R exp(t.V)(D_L exp(-t.V))exp(-i\Omega))$$

$$+exp(i\Omega)D_R^T\epsilon D_R D_L(exp(-i\Omega))$$

The second term is zero since $D_R^T\epsilon D_R.exp(-i\Omega) = 0$ because $D_R.exp(-i\Omega) = 0$ and $[D_R^T\epsilon D_R, D_L]$ is proportional to $\gamma^\mu\partial_\mu D_R$. Thus under gauge transformations, W_L transforms as

$$W_L \to exp(i\Omega)W_L.exp(-i\Omega)$$

which is consistent with the fact that since W_L is left Chiral, its gauge transform should also be left Chiral. It is then easy to see that the quantity

$$Tr[W_L^T\epsilon W_L]_F = [W_L^{AT}\epsilon W_L^A]_F$$

is gauge invariant, Lorentz invariant and supersymmetry invariant where the subscript F denotes the coefficient of $\theta_L^T\epsilon\theta_L$ in the expansion of a Chiral superfield. The matter superfield Φ is left Chiral and has a scalar field component, a left handed Dirac field component and an auxiliary F component. The field

$L_1 = [\Phi^* exp(-t.V)\Phi]_D$ is a supersymmetric Lagrangian that is also gauge invariant since under a gauge transformation,

$$exp(-t.V) \to exp(-i\Omega^*)exp(t.V)exp(i\Omega),$$

while on the other hand,

$$\Phi \to exp(-i\Omega)\Phi, \Phi^* \to \Phi^* exp(i\Omega^*)$$

L_1 describes the scalar field, the Dirac field and their interactions with gauge field in a way that generalizes the interaction of the Dirac field with the gauge fields like the electromagnetic field and more generally, with non-Abelian gauge fields. If the left Chiral superpotential field $f(\Phi)$ is also taken into account by using its F-term, then we obtain a Lagrangian for matter interacting with gauge fields that is supersymmetry, Lorentz and gauge invariant. This Lagrangian can be written down easily and we leave it as an exercise.

3.11 Quantization of the Maxwell and Dirac field in a background curved metric of space-time

Let Γ_μ denote the spinor connection of the gravitational field. If V_a^μ is the tetrad field of the metric, ie,

$$g^{\mu\nu} = \eta^{ab} V_a^\mu V_b^\nu$$

where η is the Minkowski metric, then it is known (see for example, [a] Steven Weinberg, "Gravitation and Cosmology: Principles and Applications of the General Theory of Relativity", Wiley, or [b] Steven Weinberg, "The quantum theory of fields, vol.III, Supersymmetry", Cambridge University Press) that

$$\Gamma_\mu = (1/2) V_{a\nu:\mu} V_b^\nu [\gamma^a, \gamma^b]$$

and that Dirac's equation that is diffeomorphic as well as locally Lorentz invariant is given by

$$[\gamma^a V_a^\mu (i\partial_\mu + eA_\mu + i\Gamma_\mu) - m_0]\psi = 0$$

Local Lorentz invariance is checked by noting that if $\Lambda(x)$ is a local Loretnz transformation (ie, a space-time dependent Lorentz transformation matrix w.r.t. the Minkowski metric η, ie, $\Lambda(x)^T \eta \Lambda(x) = \eta$), and if $D(.)$ denotes Dirac's spinor representation of the Lorentz group, then

$$V_a^\mu(x) D(\Lambda(x)) \gamma^a (\partial_\mu + \Gamma_\mu(x)) D(\Lambda(x))^{-1} =$$

$$= V_a^\mu D(\Lambda) \gamma^a D(\Lambda)^{-1} D(\Lambda)(\partial_\mu + \Gamma_\mu(x)) D(\Lambda(x))^{-1}$$

$$= V_a^\mu \Lambda_b^a \gamma^b D(\Lambda)(\partial_\mu + \Gamma_\mu(x)) D(\Lambda(x))^{-1}$$

$$= V_a^{\mu'}(x)\gamma^a(\partial_\mu + \Gamma'_\mu(x))$$

where $\Gamma'_\mu(x)$ is obtained from $\Gamma_\mu(x)$ by replacing $V_a^\mu(x)$ by $V_a^{\mu'}(x)$ with

$$V_a^{\mu'}(x) = \Lambda_a^b(x)V_b^\mu(x)$$

This is proved by assuming $\Lambda(x) = I + \omega(x)$ to be an infinitesimal Local Lorentz transformation. Note that $i\Gamma_\mu(x)$ is not a Hermitian matrix since $[\gamma^a, \gamma^b]$ is not skew-Hermitian for all $a, b = 0, 1, 2, 3$.

Remark: $\gamma^{0*} = \gamma^0, \gamma^{a*} = -\gamma^a, a = 1, 2, 3$ and hence $[\gamma^0, \gamma^a]$ is Hermitian for $a = 1, 2, 3$ while $[\gamma^a, \gamma^b]$ is skew-Hermitian for $a, b = 1, 2, 3$.

However, it can be shown using integration by parts that the action functional for the Dirac field in curved space-time defined by

$$S[\psi, \psi^*] = i\int \psi^*(x)\gamma^0[\gamma^a V_a^\mu(x)(\partial_\mu + \Gamma_\mu(x)) - m_0]\psi(x)\sqrt{-g(x)}d^4x$$

is real and apart from being locally Lorentz invariant, it is also diffeomorphic invariant.

The electron propagator in a curved background metric: The electron propagator is clearly given by the formal operator theoretic expression

$$S_e = (V_a^\mu \gamma^a(i\partial_\mu + i\Gamma_\mu))^{-1}$$

More specifically, if $S_e(x, y)$ is the position space kernel representation of the electron propagator, then

$$[i\gamma^a V_a^\mu(x)(\partial_\mu + \Gamma_\mu(x)) - m_0]S_e(x, y) = \delta^4(x - y)$$

We shall obtain an approximate solution to the electron propagator using perturbation theory. Let $S_{e0}(x - y)$ denote the unperturbed electron propagator. Then,

$$[i\gamma^a\partial_a - m_0]S_{e0}(x - y) = \delta^4(x - y)$$

with solution

$$S_{e0}(x - y) = [i\gamma^a\partial_a - m_0]^{-1}\delta^4(x - y)$$

or equivalently, using Fourier transforms,

$$S_{e0}(x - y) = (2\pi)^{-4}\int (i\gamma^a p_a - m + i0)^{-1} exp(ip.(x - y))d^4p$$

Let $\delta S_e(x, y)$ denote the correction to the electron propagator upto first order perturbation theory caused by the gravitational terms. We then have upto first order in the gravitational metric perturbations from the flat space-time Minkowski metric,

$$V_a^\mu = \delta_a^\mu + \delta V_a^\mu(x)$$

and hence upto the first order,

$$\Gamma_\mu = \delta\Gamma_\mu = (1/2)[\gamma^a, \gamma^b]\delta_a^\nu[\delta V_{b\nu,\mu} - \Gamma_{\nu\mu}^\rho \eta_{b\rho}]$$

$$= (1/2)[\gamma^a, \gamma^b][\delta V_{ba,\mu} - \Gamma_{ba\mu}]$$

Note that upto first order,

$$g_{\mu\nu} = \eta_{ab} V_\mu^a V_\nu^b = \eta_{ab}(\delta_\mu^a + \delta V_\mu^a)(\delta_\nu^b + \delta V_\nu^b)$$

$$= \eta_{\mu\nu} + \eta_{ab}\delta_\mu^a \delta V_\nu^b + \eta_{ab}\delta_\nu^b \delta V_\mu^a$$

$$= \eta_{\mu\nu} + 2\delta V_{\mu\nu}$$

where the tetrad $V_{a\mu}$ has been chosen to be symmetric in its two indices. In fact, if the metric is expressed as

$$g_{\mu\nu} = \eta_{\mu\nu} + \delta g_{\mu\nu}$$

then we can choose

$$V_{\mu\nu} = \delta g_{\mu\nu}/2$$

Now writing the differential equation satisfied by the propagator as

$$[i\gamma^a(\delta_a^\mu + \delta V_a^\mu(x))(\partial_\mu + \delta\Gamma_\mu(x)) - m_0][S_{e0}(x-y) + \delta S_e(x,y)) = \delta^4(x-y)$$

we get on equating first order terms,

$$[i\gamma^\mu\partial_\mu - m_0]\delta S_e(x,y) + i\gamma^a \delta V_a^\mu(x)\partial_\mu S_{e0}(x-y) + i\gamma^\mu \delta\Gamma_\mu(x) S_{e0}(x-y) = 0$$

from which we deduce the following formula for the first order propagator correction:

$$\delta S_e(x,y) = -i\int S_{e0}(x-z)[\gamma^a \delta V_a^\mu(z)\partial_\mu S_{e0}(z-y) - \gamma^\mu \delta\Gamma_\mu(z) S_{e0}(z-y)]d^4z$$

$$= -i\int S_{e0}(x-z)\gamma^a[V_a^\mu(z)\partial_\mu S_{e0}(z-y) + \delta\Gamma_a(z)S_{e0}(z-y)]d^4z$$

In order to relate all this to quantum antennas, we must also calculate the Dirac four current density in curved space time. Consider

$$\psi^*\gamma^0[(V_a^\mu\gamma^a(i\partial_\mu + i\Gamma_\mu) - m_0]\psi = 0$$

or equivalently,

$$\psi^*[V_a^\mu\alpha^a(i\partial_\mu + i\Gamma_\mu) - m_0\beta]\psi = 0$$

or equivalently,

$$i\partial_\mu[\psi^* V_a^\mu \alpha^a\psi] - i\partial_\mu[\psi^* V_a^\mu]\alpha^a\psi$$
$$+\psi^*[iV_a^\mu\alpha^a\Gamma_\mu - m_0\beta]\psi = 0$$

Taking the conjugate of this equation gives

$$-i\partial_\mu[\psi^* V_a^\mu\alpha^a\psi] + i\psi^*\alpha^a\partial_\mu[V_a^\mu\psi]$$
$$+\psi^*[-iV_a^\mu\Gamma_\mu^*\alpha^a - m_0\beta]\psi = 0$$

3.12 Relationship between the electron self energy and the electron propagator

Consider an electron bound to its nucleus. Let E_n denote its energy eigenvalue corresponding to the eigenfunction $u_n(\mathbf{x})$. Then, the propagator of the electron in the second quantized picture can be expressed as

$$S(x,y) = S(t,\mathbf{x}|(t',\mathbf{y}) = \theta(t-t')\sum_n u_n(\mathbf{x})\bar{u}_n(\mathbf{y})exp(-iE_n(t-t'))$$

To check this, we prove that S satisfies the propagator differential equation

$$(i\partial_t - H)S(t,\mathbf{x}|t',\mathbf{y}) = i\delta^4(x-y), x = (t,\mathbf{x}), y = (t',\mathbf{y})$$

This is proved using the identities

$$\theta'(t-t') = \delta(t-t'), \sum_n u_n(\mathbf{x})\bar{u}_n(\mathbf{y}) = \delta^3(\mathbf{x}-\mathbf{y})$$

In the frequency/energy domain, the propagator is given by

$$S(\mathbf{x},\mathbf{y}|E) = \int_{\mathbb{R}} S(t,\mathbf{x}|0,\mathbf{y})exp(iE(t-t'))dt =$$

$$i\sum_n u_n(\mathbf{x})\bar{u}_n(\mathbf{y})/(E-E_n)$$

or equivalently, in operator theoretic notation,

$$S(E) = i\sum_n |u_n><u_n|/(E-E_n), <u_n|u_m>= \delta_{n,m}$$

In particular,

$$\int \bar{u}_n(\mathbf{x})S(\mathbf{x},\mathbf{y}|E)d^3x =$$

$$i\bar{u}_n(y)/(E-E_n)$$

Then the change in the propagator caused by radiative effects in which the energy levels get perturbed by δE_n and correspondingly, the stationary state eigenfunctions get perturbed by $\delta u_n(\mathbf{x})$ is given by

$$-i\delta S(\mathbf{x},\mathbf{y}|E) = \sum_n[\delta u_n(\mathbf{x})\bar{u}_n(\mathbf{y})+u_n(\mathbf{x})\delta u_n(\mathbf{y})]/(E-E_n)+\sum_n u_n(\mathbf{x})\bar{u}_n(\mathbf{y})\delta E_n/(E-E_n)^2$$

It follows then that on writing the one loop radiative correction to the electron propagator as

$$\delta S = S.\Sigma.S$$

that

$$-i<u_n|\delta S|u_n>= -i<u_n|S\Sigma.S|u_n>= -i<u_n|\Sigma|u_n>/(E-E_n)^2$$

on the one hand while on the other,

$$-i < u_n|\delta S|u_n >= \delta E_n/(E - E_n)^2$$

where we have used the orthogonality relation

$$< \delta u_n|u_n >= 0$$

since both u_n and $u_n + \delta u_n$ are normalized, ie have unit norm. This gives us the fundamental relation between the change in the electron propagator caused by one loop radiative corrections and the shift in the electron energy as

$$\delta E_n = -i < u_n|\Sigma|u_n >$$

This is the extra energy gained by the electron due to propagator corrections coming from radiative as well as gravitational effects.

3.13 Electron self energy corrections induced by quantum gravitational effects

For the free gravitational field, let

$$< 0|T(\delta\Gamma_\mu(x)\delta\Gamma_\nu(y))|0 >= D_\Gamma(x, y)$$

This can be viewed as some sort of propagator for the free quantum gravitational field. The wave equation satisfied by the Dirac field in the presence of quantum gravitational effects is given by

$$[i\gamma^a(\delta^\mu_a + \delta V^\mu_a(x))(\partial_\mu + \delta\Gamma_\mu(x)) - m_0][\psi^{(0)}(x) + \delta\psi(x)] = 0$$

$\psi^{(0)}$ is the free electron-positron wave operator field. It satisfies the zeroth order perturbation equation:

$$[i\gamma^\mu\partial_\mu - m_0]\psi^{(0)} = 0$$

and its solution is expressible as a superposition of the electron annihilation operators and the positron creation operators in momentum space. $\delta\psi$ is the correction to the free Dirac field caused by gravitational effects upto first order. It satisfies the first order perturbation equation:

$$[i\gamma^\mu\partial_\mu - m_0]\delta\psi(x)$$

$$+i\gamma^a\delta V^\mu_a(x)\partial_\mu\psi^{(0)}(x) + i\gamma^\mu\delta\Gamma_\mu(x)\psi^{(0)}(x) = 0$$

ie $\delta\psi(x)$ satisfies the same differential equation as the first order perturbation $\delta S_e(x, y)$ in the electron propagator and its solution is given by

$$\delta\psi(x) = -i\int S_{e0}(x - y)[\gamma^a\delta V^\mu_a(y)\partial_\mu\psi^{(0)}(y) + \gamma^\mu\delta\Gamma_\mu(y)\psi^{(0)}(y)]d^4y$$

and hence the approximate corrected electron propagator upto linear orders in the graviton propagator is given by

$$S_e(x,y) =< 0|T\{(\psi^{(0)}(x)+\delta\psi(x)).(\psi^{(0)}(y)+\delta\psi(y))^*\}|0>$$

$$= S_{e0}(x-y)+ < 0|T\{\psi^{(0)}(x).\delta\psi(y)^*\}|0> + < 0|T\{\delta\psi(x).\psi^{(0)*}(y)\}|0>$$
$$+ < 0|T\{\delta\psi(x).\delta\psi(y)^*\}|0>$$
$$= S_{e0}(x-y)+ < 0|T\{\delta\psi(x).\delta\psi(y)^*\}|0>$$

with

$$< 0|T\{\delta\psi(x).\delta\psi(y)^*\}|0>=$$

$$\int S_{e0}(x-u)[\gamma^a\delta V_a^\mu(u)\partial_\mu\psi^{(0)}(u)+\gamma^\mu\delta\Gamma_\mu(u)\psi^{(0)}(u)].[(\delta V_b^\nu(v)\partial_\nu\psi^{(0)}(v)^*\gamma^{b*}$$
$$+\psi^{(0)}(v)^*\delta\Gamma_\nu(v)^*\gamma^{\nu*}]S_{e0}(y-v)^*d^4ud^4v$$

Chapter 4

Miscellaneous Problems

4.1 A problem in robotics

Let $\mathbf{R}_k(t), k = 1, 2, ..., N$ denote the positions of moving point objects which have to be picked up by a robot. More generally, instead of point objects, we can assume extended rigid bodies moving on the ground or in space which have to be picked up by the robot. We have a camera in synchronization with the robot at the location $\mathbf{R}(t)$ (ie the position of its centre of mass). Now the camera takes pictures of the point objects and also of the robot and a digital computer calculates the distances and bearings of the images of the objects with that of the robot and accordingly generates control torques that are used to manipulate the robot so that it moves closer to one of the objects, say the mth one in succession, ie, the robot uses the error in the images of the position of the mth object and that of the robot to generate a control torque signal that eventually enables the robot to track this object and finally reduce the error in its position relative to the robot to zero and finally pick the object up. This series of jobs is performed successively on the different objects so that finally all the objects are picked up. The mathematical details of formulating an algorithm are based on the gradient descent algorithm and could be described as follows: Let $I(\mathbf{R}_k(t), x, y)$ denote the image field on the camera screen generated by the kth object at time t and let $I(\mathbf{R}(t), \mathbf{q}(t), x, y)$ be the image field on the camera screen generated by the robot at time t whose centre of mass is located at $\mathbf{R}(t)$ and whose link angles relative to a given direction are denoted by $\mathbf{q}(t)$. The computer calculates the error energy

$$E_k(t, \mathbf{R}(t), \mathbf{q}(t)) = \int_{screen} (I(\mathbf{R}_k(t), x, y) - I(\mathbf{R}(t), \mathbf{q}(t), x, y))^2 dxdy$$

and then the computer generates the following algorithm for moving the robot using a force and torque that causes the robot's location and link angles respectively to change after a small time δt to

$$\mathbf{R}(t) + \delta\mathbf{R}(t), \mathbf{q}(t) + \delta\mathbf{q}(t)$$

where

$$\delta\mathbf{R}(t) = -\mu.\delta t\nabla_{\mathbf{R}(t)}E_k(t,\mathbf{R}(t),\mathbf{q}(t)),$$

$$\delta\mathbf{q}(t) = -\mu.\delta t\nabla_{\mathbf{q}(t)}E_k(t,\mathbf{R}(t),\mathbf{q}(t))$$

Our aim will be to generalize this model to the tracking and picking up rigid bodies and even non-rigid extended bodies. The force and torque generation mechanism are based on Newtonian mechanics:

$$\mathbf{F}(t) = M\mathbf{R}''(t), \tau(t) = J(\mathbf{q}(t))\mathbf{q}''(t) + N(\mathbf{q}(t),\mathbf{q}'(t))$$

where M, J are respectively the robot mass and its mass moment of inertia matrix and $N(\mathbf{q}(t),\mathbf{q}'(t))$ consists of centrifugal, coriolis, frictional and gravitational potential contributions to the computed torques. In practice, these control forces and torques are generated by discretizing the time derivatives with a time step of δt:

$$\delta\mathbf{q}''(t) \approx (\delta\mathbf{q}(t) - 2\delta\mathbf{q}(t-\delta t) + \delta\mathbf{q}(t-2\delta t))/(\delta t)^2$$

$$\mathbf{R}''(t) = (\mathbf{R}(t) - 2\mathbf{R}(t-\delta t) + \mathbf{R}(t-2\delta t))/(\delta t)^2$$

We propose in our project to do a noise analysis of this algorithm based on Varadhan's large deviation theory. We also propose to do a robustness analysis of this problem based on how sensitive is the tracking error energy to errors induced in camera imaging and also to errors induced in the digital computer due to finite register effects while computing the control forces and torques. This work is to be regarded as an extension of a paper [1] based on the gradient search algorithm. [1] does not consider a mathematical analysis of noise effects on the algorithm. We propose to do such an analysis by adding WGN to the rhs of the gradient algorithm thereby resulting in a nonlinear stochastic difference equation and we shall use the standard techniques based on mean and variance propagation to analyze these effects [6]. The gradient algorithm for developing the computed force and torques are based on the gradient algorithm which take into account only the instantaneous error. In our project, we shall also be considering generalizations of this based on past error history.

4.2 More on root space decomposition of a semisim-ple Lie algebra

Let $\mathfrak{g}$ be a semisimple Lie algebra and let $\mathfrak{h}$ be a Cartan subalgebra of it. Then, $\mathfrak{h}$ is Abelian, $ad(\mathfrak{h})$ is an Abelian family of semisimple linear operators on $\mathfrak{g}$ and hence can be simultaneously diagonalized. Thus we get the root space decomposition of $\mathfrak{g}$ as

$$\mathfrak{g} = \mathfrak{h} \oplus \bigoplus_{\alpha\in\Delta} \mathfrak{g}_\alpha$$

where Δ is a finite subset of $\mathfrak{h}^*$, none of which is zero (The zero eigenspace of $ad(\mathfrak{h})$ is precisely $\mathfrak{h}$ itself since $\mathfrak{h}$ is maximal Abelian) and

$$ad(H)(X) = [H, X] = \alpha(H)X, iffX \in \mathfrak{g}_\alpha \forall H \in \mathfrak{h}$$

In other words, for any $\alpha \in \Delta$, $\mathfrak{g}_\alpha$ is the eigenspace of $ad(H)$ corresponding to the eigenvalue $\alpha(H)$ for every $H \in \mathfrak{h}$. An element of Δ is called a root. We now claim that

$$\Delta = -\Delta$$

In fact, we have that for any $X \in \mathfrak{g}_\alpha, Y \in \mathfrak{g}_\beta$ the identity

$$B([H, X], Y) = -B(X, [H, Y]), H \in \mathfrak{h}$$

and hence

$$(\alpha(H) + \beta(H))B(X, Y) = 0$$

Further,

$$\alpha(H')B(H, X) = B(H, [H', X]) = -B([H, H'], X) = 0, H, H' \in \mathfrak{h}$$

and since $\alpha \in \mathfrak{h}^*$ is nonzero, it follows that

$$B(H, X) = 0 \forall H \in \mathfrak{h}$$

In other words, we have proved two things: One, that $\mathfrak{h} \perp \mathfrak{g}_\alpha \forall \alpha \in \Delta$ and two that $\beta \neq -\alpha$ implies that $\mathfrak{g}_\alpha \perp \mathfrak{g}_\beta$. From these two and the above root space decomposition of $\mathfrak{g}$, it easily follows that if there is an $\alpha \in \Delta$ such that $-\alpha \notin \Delta$, then $g_\alpha \perp \mathfrak{g}$ which contradicts the non-degeneracy of $B(.,.)$. This proves that $\Delta = -\Delta$. Next, we prove that $dim\mathfrak{g}_\alpha = 1 \forall \alpha \in \Delta$ and $\mathfrak{g}_{k\alpha} = 0, k > 1\alpha \in \Delta$. Indeed consider the subspace V of $\mathfrak{g}$ defined by

$$V = span\{Y\} \oplus \mathfrak{h}\mathfrak{g}_\alpha \oplus \mathfrak{g}_{2\alpha} \oplus ... \oplus \mathfrak{g}_{k\alpha} \oplus ...$$

the series terminating after a finite number of steps since $\mathfrak{g}$ and hence V are finite dimensional. Here, $Y \in \mathfrak{g}_{-\alpha}$ is any non-zero element. Note that for any $\alpha, \beta \in \Delta$, we have

$$[\mathfrak{g}_\alpha, \mathfrak{g}_\beta] \subset \mathfrak{g}_{\alpha+\beta}$$

and

$$[\mathfrak{g}_\alpha, \mathfrak{g}_{-\alpha}] \subset \mathfrak{h} \forall \alpha, \beta \in \Delta$$

the first follows by Jacobi's identity and the second by Jacobi's identity and maximal Abelian property of $\mathfrak{h}$. Now choose $0 \neq X \in \mathfrak{g}_\alpha$ and Then $ad(X)$ leaves V invariant while $ad(Y)$ also leaves V invariant. Then, $ad(H)$ with $H = [X, Y] \in \mathfrak{h}$ also leaves V invariant. But

$$ad(H) = [ad(X), ad(Y)]$$

and hence $Tr(ad(H)|_V) = 0$. Thus we get

$$0 = -\alpha(H) + \alpha(H)dim(\mathfrak{g}_\alpha) + 2\alpha(H)dim(\mathfrak{g}_{2\alpha}) + ...$$

Choosing $H \in \mathfrak{h}$ so that $\alpha(H) \neq 0$, we get

$$0 = -1 + dim(\mathfrak{g}_\alpha) + 2dim(\mathfrak{g}_{2\alpha}) + ...$$

and this easily results in

$$dim(\mathfrak{g}_\alpha) = 1, dim(\mathfrak{g}_{k\alpha}) = 0, k \geq 2$$

Now let $\alpha, \beta \in \Delta$. We look at the root spaces $\mathfrak{g}_{\alpha+k\beta}, k = -p, -p+1, ..., q-1, q$ where $\{\alpha + k\beta : k = -p, -p+1, ..., q-1, q\}$ is a maximal chain, ie, $\alpha + k\beta, k = -p, p+1, ..., q-1, q$ are all roots, ie, elements of Δ but $\alpha + (q+1)\beta$ and $\alpha - (p+1)\beta$ and $\alpha + (q+1)\beta$ are not roots. By maximality, it is clear that no other chain of this sort can overlap with this chain. Note that we have already proved that $dim(\mathfrak{g}_{\alpha+k\beta}) = 1, k = -p, ..., q$. Our immediate aim is to show that $V = \bigoplus_{k=-p}^{q} \mathfrak{g}_{\alpha+k\beta}$ is a vector space such that the restriction of the $sl(2, \mathbb{C})$ Lie algebra generated by $H_\alpha, X_\alpha, X_{-\alpha}\}$ has its adjoint representation restricted to V an irreducible representation of $sl(2, \mathbb{C})$. Here, $X_\alpha \in \mathfrak{g}_\alpha$ and $X_{-\alpha} \in \mathfrak{g}_{-\alpha}$ are nonzero elements and their normalizations are chosen so that $B(X_\alpha, X_{-\alpha}) = 1$. Then, we have

$$[X_\alpha, X_{-\alpha}] = c\bar{H}_\alpha$$

for some $\bar{H}_\alpha \in \mathfrak{h}$ and hence,

$$cB(\bar{H}_\alpha, H) = B([X_\alpha, X_{-\alpha}], H) = B(X_\alpha, [X_{-\alpha}, H]) =$$

$$\alpha(H)B(X_\alpha, X_{-\alpha}), H \in \mathfrak{h}$$

We choose $\bar{H}_\alpha$ so that

$$B(\bar{H}_\alpha, H) = \alpha(H), H \in \mathfrak{h}$$

This is possible in view of the above equation by taking

$$c = B(X_\alpha, X_{-\alpha})$$

Note that since B is non-singular on $\mathfrak{g}$, and X_α is orthogonal to both $\mathfrak{h}$ and to $\mathfrak{g}_\beta \forall \beta \neq -\alpha$, and since $X_{-\alpha}$ is a non-zero element of the one dimensional vector space $\mathfrak{g}_{-\alpha}$, it follows that $B(X_\alpha, X_{-\alpha}) \neq 0$. Thus with the above choice of c, we get that

$$[X_\alpha, X_{-\alpha}] = B(X_\alpha, X_{-\alpha})\bar{H}_\alpha$$

Now we define

$$H_\alpha = B(X_\alpha, X_{-\alpha})\bar{H}_\alpha$$

and obtain

$$[X_\alpha, X_{-\alpha}] = H_\alpha$$

and then,

$$B(H_\alpha, H_\alpha) = B(H_\alpha, [X_\alpha, X_{-\alpha}]) =$$

$$B([H_\alpha, X_\alpha], X_{-\alpha}) =$$
$$\alpha(H_\alpha)B(X_\alpha, X_{-\alpha})$$

If we choose the normalizations of $X_\alpha, X_{-\alpha}$ so that

$$B(X_\alpha, X_{-\alpha}) = 2/B(\bar{H}_\alpha, \bar{H}_\alpha) = 2/\alpha(\bar{H}_\alpha)$$

then we get

$$H_\alpha = 2\bar{H}_\alpha/\alpha(\bar{H}_\alpha) = 2\bar{H}_\alpha/B(\bar{H}_\alpha, \bar{H}_\alpha)$$

and then we get

$$[H_\alpha, X_\alpha] = \alpha(H_\alpha)X_\alpha = 2X_\alpha,$$
$$[H_\alpha, X_{-\alpha}] = -\alpha(H_\alpha)X_{-\alpha} = -2X_{-\alpha}$$

and as before,

$$[X_\alpha, X_{-\alpha}] = H_\alpha$$

In other words, the triplet $\{H_\alpha, X_\alpha, X_{-\alpha}\}$ form a standard set of generators of an $sl(2, \mathbb{C})$ Lie algebra.

4.3 A project proposal for developing an experimental setup for transmitting quantum states over a channel in the presence of an eavesdropper

In quantum computation and information theory, it is by now a well established fact that a qubit state and more generally d-qubit state (ie a pure state in $\mathbb{C}^{2^d}$) can be transmitted over a channel from A to B by transmitting just $2d$ classical bits provided that A and B share a maximally entangled state, ie, a state of the form $d^{-1/2}\sum_{k=0}^{d-1}|k, k>$. The idea is simply to append this state to the maximally entangled state at $A's$ end, then perform a unitary transformation on the total $2d$-qubit state of A, perform a measurement at $A's$ end thereby causing $B's$ state to collapse to one of $2d$ possible d-qubit states. When A then reports to B about his measurement outcome via $2d$ classical bits, B is able to apply an appropriate unitary gate at his end to recover the original state that A had intended to transmit. In quantum information theory, another important problem is the Cq problem in which A wishes to transmit classical information over a quantum channel by encoding his classical bits in the form of quantum states. Thus, if $A's$ classical information source is the alphabet $A = \{1, 2, ..., a\}$, with the alphabet k occurring with probability $p(k)$, then the total information contained in this source that A wishes to transmit is $H(A) = H(p) = -\sum_{x=1}^{a} p(x).log(p(x))$. A encodes the alphabet $x \in A$ in the form of a density matrix $\rho(x)$ (ie, a mixed state in a finite dimensional Hilbert space $\mathcal{H}$), and transmits this state over the channel assumed to be noiseless.

The state received by B is then $\rho(x)$ and the average state received by B is $\bar{\rho} = \sum_{x \in A} p(x)\rho(x)$. It is natural to expect that the total information that A has transmitted to B must be given by

$$I_p(A, B) = H(B) - H(B|A) = H(\bar{\rho}) - \sum_{x \in A} p(x)H(\rho(x))$$

where

$$H(W) = -Tr(W.log(W))$$

is the Von-Neumann entropy of the state W. In order for this to be a meaningful measure of the information transmitted, we can ask the following question: Suppose A encodes a string of his source alphabets $\mathbf{x} = (x(1), ..., x(n)) \in A^n$ into the state $\rho(\mathbf{x}) = \rho(x(1)) \otimes ... \otimes \rho(x(n))$ in the tensor product Hilbert space $\mathcal{H}^{\otimes n}$, then can he choose M_n such distinct sequences $\mathbf{x}_1, ..., \mathbf{x}_M$ such that (a) these sequences are all typical for A' source w.r.t to the probability distribution $p^n(\mathbf{x}) = p(x(1))...p(x(n))$, (b) There exist positive "Detection Operators" $\mathbf{D_1}, ..., \mathbf{D_{M_n}}$ for B in the Hilbert space $\mathcal{H}^{\otimes n}$ such that for any $\epsilon > 0$ with n sufficiently large, one has

$$\mathbf{D}_1 + ... + \mathbf{D}_{M_n} \leq I,$$

$$Tr(\rho(\mathbf{x}_k)\mathbf{D}_k) > 1 - \epsilon, k = 1, 2, ..., M_n$$

and

$$Tr(D_k) \leq Tr(E(\mathbf{x}_k, n, \delta)), k = 1, 2, ..., M_n$$

where $E(\mathbf{x}, n, \delta)$ is a δ-typical projection on $\mathcal{H}^{\otimes n}$ corresponding to the situation when $\mathbf{x}$ is a d-typical sequence for $A's$ source. These requirements amount to saying that the detection operator $\mathbf{D}_k$ of B does not have too large a dimension so as to "leak" into another sequence $\mathbf{x}_j, j \neq k$ when $\mathbf{x}_k$ is transmitted, and further, that with a large probability, $B's$ decision on what sequence A had transmitted is correct when he uses his detection operators. Then the question is that if M_n is maximal subject to these requirements, so that the rate of reliable transmission of information (ie, with error probability smaller than ϵ), is $\frac{log(M_n)}{n}$, then $limlog(M_n)/n = sup_p I_p(A, B)$. In other words, the maximum rate of reliable transmission of information on a Cq channel is precisely the Cq channel capacity defined by

$$C = sup_p I_p(A, B) = sup_p(H(\sum_x p(x)\rho(x)) - \sum_x p(x)H(\rho(x)))$$

Our project will involve verifying this capacity formula by preparing the states $\rho(x), x \in A$ using lasers and ions, so that by shining the laser on an ion, we can generate excited ion states used for transmission. In other words, one of the primary objectives of our experimental setup will be to prepare a large class of quantum states using the quantum electromagnetic field generated by a laser interacting with ions. If the ion and the laser field start in an initial state $|k, \phi(u)>$ where $|k>$ represents a stationary state of the ion and $|\phi(u)>$

a coherent state of the laser, then after interacting with each other for a time duration T, the final state of the ion and the laser field will be the pure state $U(T)|k, \phi(u) >$ and by partially tracing this out over the laser field state, the ion state becomes the mixed state

$$\rho_{ion}(k, u, T) = Tr_2(U(T)|k, \phi(u) >< k, \phi(u)|U(T)^*)$$

By varying k, u, T, namely, the intial state of the ion, the coherent state of the laser field and the time duration of interaction of these two, we can thus generate a host of mixed states on the Hilbert space of the ion. These mixed states can be used for transmission.

Another application of our experimental setup will be to create entangled states between three people A, B, E or more generally a mixed state ρ_{ABE} in the tensor product Hilbert space $\mathcal{H}_{ABE} = \mathcal{H}_A \otimes \mathcal{H}_B \otimes \mathcal{H}_E$ and to transmit maximal information from A to B while restricting the information transmitted from A to E to be a minimum. Let A have a classical source with alphabet A and source probability distribution $p(x), x \in A$ and let for each $x \in A$ $\rho_{BE}(x)$ be a state in $\mathcal{H}_{BE} = \mathcal{H}_B \otimes \mathcal{H}_E$. Then the information transmitted from A to B is given by

$$I(A, B) = H(\sum_x p(x)\rho_B(x)) - \sum_x p(x)H(\rho_B(x))$$

while the information transmitted from A to E is given by

$$I(A, E) = H(\sum_x p(x)\rho_E(x)) - \sum_x p(x)H(\rho_E(x))$$

where

$$\rho_B(x) = Tr_E(\rho_{BE}(x)), \rho_E(x) = Tr_B(\rho_{BE}(x))$$

The problem is to select the probability distribution $p(x), x \in A$ for $A's$ source and the states $\rho_{BE}(x), x \in A$ so that

$$I(A, B) - I(A, E)$$

is a maximum, ie, to transmit maximum information across the Cq channel from A to B while keeping the Cq information transmitted to E a minimum. Such a setup can be arranged using lasers and ion trap experiments as follows: Let A generate a current $I(t, x)$ dependent upon his source alphabet $x \in A$ and let him connect this current to a classical antenna that transmits electromagnetic waves to both B and E. The magnetic vector potential at B is then given by

$$\mathbf{A}_B(t, \xi) = \int_{S_A} G(t - s, \mathbf{R}_B + \xi, \mathbf{u})\mathbf{J}(s, \mathbf{u}, x)dsdu$$

while the magnetic vector potential at $E's$ end is given by

$$\mathbf{A}_E(t, \xi) = \int_{S_A} G(t - s, \mathbf{R}_C + \xi, \mathbf{u})\mathbf{J}(s, \mathbf{u}, x)dsdu$$

where $G(t-s, \mathbf{R}, u)$ is the standard retarded potential Green's function between the point $\mathbf{u}$ on $A's$ antenna surface the point $\mathbf{R}$ of reception of the field. $\mathbf{R}_B$ is the location of the nucleus of $B's$ atomic receiver while $\mathbf{R}_E$ is the location of the nucleus of $E's$ atomic receiver. $\mathbf{J}(t, \mathbf{u}, x)$ is the surface current density on $A's$ antenna surface S_A and it depends upon the alphabet x that A wishes to transmit and hence from basic antenna theory, it can be expressed as

$$\mathbf{J}(t, \mathbf{u}, x) = \int \mathbf{F}(t-s, \mathbf{u}) I(s, x) ds, x \in A$$

where the function $\mathbf{F}(t, \mathbf{u}), t \in \mathbb{R}, \mathbf{u} \in S_A$ depends only upon the antenna surface geometry and the point on this surface where the current source $I(t, x)$ us fed in. There is an interaction potential V_{BE} between the systems used by B and E and hence, the Schrodinger equation for the joint state ρ_{BE} when A transmits the symbol x has the form

$$i\rho'_{BE}(t) = [H_{BE}(t), \rho_{BE}(t)]$$

where

$$H_{BE}(t) = H_B(t) + H_E(t) + V_{BE}$$

with

$$H_B(t) = (\mathbf{p}_B + e\mathbf{A}_B(t, \xi_B)^2/2m - Ze^2/|\xi_B| - e\Phi_B(t, \xi_B),$$
$$H_E(t) = (\mathbf{p}_E + \mathbf{A}_E(t, \xi_E))^2/2m - Ze^2/|\xi_E| - e\Phi_E(t, \xi_E)$$

and

$$\mathbf{p}_B = -i\nabla_{\xi_B}, \mathbf{p}_E = -i\nabla_{\xi_E}$$

When A uses a quantum antenna source to transmit a quantum electromagnetic field, then the fields $\mathbf{A}_B, \mathbf{A}_E$ also become quantum fields and then the above Schrodinger equation for $\rho_{BE}(t)$ must be partially traced out over the coherent state of the bath field to obtain the "system part" of $\rho_{BE}(t)$. Note that since the surface current density $\mathbf{J}(t, \mathbf{u}, x)$ in $A's$ antenna depends upon the symbol x that A wishes to transmit, it follows that $\rho_{BE}(t) = \rho_{BE}(t, x)$ will also depend upon the symbol x and then the Cq approach mentioned above can be applied to design $A's$ antenna and his source probability distribution $p(x)$ for maximal transmission of information from A to B while keeping the information that has been leaked into $E's$ receiver at a minimum.

4.4 A problem in Lie group theory

Let H, X, Y be the standard generators of $sl(2, \mathbb{C})$, ie, $[H, X] = 2X, [H, Y] = -2Y, [X, Y] = H$. Then define for $t, x, y \in \mathbb{R}$,

$$g(t, x, y) = exp(tH + xX + yY)$$

and express

$$\partial_t g(t, x, y) = g(t, x, y)(a(1)H + a(2)X + a(3)Y),$$

$$\partial_x g(t,x,y) = g(t,x,y)(b(1)H + b(2)X + b(3)Y),$$

$$\partial_y g(t,x,y) = g(t,x,y)(c(1)H + c(2)X + c(3)Y)$$

Evaluate $a(k), b(k), c(k), k = 1, 2, 3$.

Hint: Use the the following formula for the differential of the exponential map:

$$\partial_t exp(A + tB) = exp(A + tB)(f(ad(A + tB))(B)$$

where

$$f(z) = (1 - exp(-z))/z$$

Hence, express the Haar measure on $SL(2, \mathbb{R})$ in terms of t, x, y, ie if μ is the Haar measure, then

$$d\mu(g(t,x,y)) = F(t,x,y)dtdxdy$$

and evaluate the Haar density F. For doing this, you must use the formulas obtained to express

$$g(t,x,y)H, g(t,x,y)X, g(t,x,y)Y$$

as linear combinations of $\partial_t g(t,x,y), \partial_x g(t,x,y), \partial_y g(t,x,y)$

Chapter 5

More Problems in Linear Algebra and Functional Analysis

Course titles: [A] Matrix theory, [B] Antennas and wave propagation, [C] Probability theory.

Course instructor:Harish Parthasarathy

5.1 Riesz representation theorem

[1] Prove Riesz' representation theorem in the following form: Let $X = C[0,1]$, the space of continuous functions on $[0,1]$ with sup-norm, ie,

$$\| x \| = sup(|x(t)| : t \in [0,1]\}, x \in X$$

Show that $(X, \| . \|)$ is a Banach space.

[2] Let (X, d) be a compact metric space and let $C(X)$ denote the space of all continuous functions on X. Prove that $C(X)$ under the sup norm is a Banach space. Let $\psi : C(X) \to \mathbb{R}$ be a continuous linear functional on $C(X)$. Then, show that there exists a unique signed measure μ on the Borel subsets of X such that

$$\psi(f) = \int_X f(x)d\mu(x), f \in C(X)$$

Show further that

$$\| \psi | = sup(|\psi(f)| : \| f \| \leq 1) = |\mu|(X)$$

where

$$\| f \| = sup(|f(x)| : x \in X)$$

and $|\mu|$ is the variation norm of μ, ie, corresponding to the Hahn decomposition

$$X = X_+ \cup X_-, X_+ \cap X_- = \phi$$

so that μ is non-negative on all the Borel subsets of X_+ and non-positive on all the Borel subsets of X_-, we define

$$|\mu|(E) = \mu(X_+ \cap E) - \mu(X_- \cap E)$$

5.2 Lie's theorem on solvable Lie algebras

Let $\mathfrak{g}$ be a solvable Lie algebra, ie for some finite positive integer p, we have

$$\mathcal{D}^{p+1}\mathfrak{g} = 0, \mathfrak{D}^p\mathfrak{g} \neq 0$$

where if S is any subset of $\mathfrak{g}$, then DS denotes the linear span of the elements $[X, Y], X, Y \in S$. Let ρ be any finite dimensional representation of $\mathfrak{g}$ in a complex vector space V. Then, we claim that there exists a non-zero vector $v \in V$ and a linear functional λ on $\mathfrak{g}$, ie, $\lambda \in \mathfrak{g}^*$ such that

$$\rho(X)v = \lambda(X)v, \forall X \in \mathfrak{g}$$

We prove this theorem by induction. $\mathfrak{g}$ is solvable, it is clear that $D\mathfrak{g}$ is a proper subspace of $\mathfrak{g}$ (In fact $D\mathfrak{g}$ is an ideal in $\mathfrak{g}$). Thus, we can choose a subspace $\mathfrak{h}$ of $\mathfrak{g}$ such that $dim\mathfrak{h} = dim\mathfrak{g} - 1$, or equivalently, $dim(\mathfrak{g}/\mathfrak{h}) = 1$ and such that

$$D\mathfrak{g} \subset \mathfrak{h} \subset \mathfrak{g}$$

It is clear that $\mathfrak{h}$ is an ideal in $\mathfrak{g}$ since

$$[X, \mathfrak{h}] \subset D\mathfrak{g} \subset \mathfrak{h}, X \in \mathfrak{g}$$

In particular,

$$D\mathfrak{h} \subset D\mathfrak{g} \subset \mathfrak{h}$$

or in other words, that $\mathfrak{h}$ is a proper Lie subalgebra of $\mathfrak{g}$ having codimension one in $\mathfrak{g}$. Hence, ρ restricted to $\mathfrak{h}$ is a representation. By the induction hypothesis (induction is on $dim\mathfrak{g}$), it follows that there is a nonzero $w_0 \in V$ such that

$$\rho(Y)w_0 = \lambda(Y)w_0, \forall Y \in \mathfrak{h}$$

for some $\lambda \in \mathfrak{h}^*$. Now since $\mathfrak{h}$ has codimension one in $\mathfrak{g}$, it is clear that we can choose an $X_0 \in \mathfrak{g}, X_0 \notin \mathfrak{h}$ and then it follows that

$$\mathfrak{g} = \{cX_0 + Y : c \in \mathbb{C}, Y \in \mathfrak{h}\} =< X_0 > +\mathfrak{h}$$

Now let q be the smallest non-negative integer for which $\{\rho(X_0)^k w_0 : 0 \leq k \leq q\}$ forms a linearly independent set. Then, $\rho(X_0)^{q+1}w_0$ is linearly dependent on $\{\rho(X_0)^k w_0 : 0 \leq k \leq q\}$ and by successive application of $\rho(X_0)$, it follows that for any $m > q$, $\rho(X_0)^m w_0$ is also linearly dependent upon $\{\rho(X_0)^k w_0 : 0 \leq k \leq q\}$. Define the subspaces

$$W_m = span\{w_0, w_1, ..., w_m\}, m = 0, 1, ..., q, w_k = \rho(X_0)^k w_0, k = 0, 1, ..., q$$

Since then $\rho(X_0)w_k = w_{k+1}, k = 0, 1, ..., q-1$ and $\rho(X_0)w_q \in W_q$, it follows that W_q is $\rho(X_0)$ invariant. We have further,

$$\rho(Y)\rho(X_0)w_k = [\rho([Y, X_0]) + \rho(X_0)\rho(Y)]w_k, Y \in \mathfrak{h} - - - (1)$$

Now, $\mathfrak{h}$ is an ideal in $\mathfrak{g}$ and hence, if we assume that the proposition P_k defined by

$$\rho(Y)w_k = \lambda(Y)w_k mod W_{k-1}, Y \in \mathfrak{h}$$

is valid for some k, then P_0 is true with $W_{-1} = 0$ and P_k implies in view of (1) that

$$\rho(Y)w_{k+1} = \lambda([Y, X_0])w_k + \lambda(Y)w_{k+1}, Y \in \mathfrak{h} - - - (2)$$

so that P_{k+1} is also true. Note that since $[Y, X_0] \in \mathfrak{h}, Y \in \mathfrak{h}$ since $\mathfrak{h}$ is an ideal. Thus, by induction, it follows that P_k is true for $k = 0, 1, ..., q$, ie,

$$\rho(Y)w_k = \lambda(Y)w_k mod W_{k-1}, k = 0, 1, ..., q \ Y \in \mathfrak{h}$$

In particular, W_q is $\rho(\mathfrak{h})$-invariant and further,

$$Tr(\rho(Y)|_{W_q}) = (q+1)\lambda(Y), Y \in \mathfrak{h} - - - (3)$$

Now, we have seen that W_q is $\rho(X_0)$ invariant and also $\rho(Y)$ invariant and hence, since $[Y, X_0] \in \mathfrak{h}$, we have

$$Tr(\rho([Y, X_0])|_{W_q}) = Tr([\rho(Y), \rho(X_0)]|_{W_q}) = 0, Y \in \mathfrak{h}$$

Thus from (3), with $[Y, X_0]$ in place of Y, we have

$$\lambda([Y, X_0]) = 0, Y \in \mathfrak{h}$$

and hence, we get from (2) that

$$\rho(Y)w_{k+1} = \lambda(Y)w_{k+1}, k = 0, 1, ..., q-1, Y \in \mathfrak{h}$$

and hence

$$\rho(Y)w_k = \lambda(Y)w_k, k = 0, 1, ..., q, Y \in \mathfrak{h}$$

In other words, for any $Y \in \mathfrak{h}$ $\rho(Y)$ acts on W_q as multiplication by $\lambda(Y)$ times the identity operator. Further, since W_q is $\rho(X_0)$-invariant, and the field is the $\mathbb{C}$ which is algebraically closed, we can choose a nonzero vector $v \in W_q$ such that

$$\rho(X_0)v = c_1 v$$

for some complex number c_1. Now, we extend the domain of definition of λ from $\mathfrak{h}^*$ to the whole of $\mathfrak{g}$ by setting

$$\lambda(cX_0 + Y) = cc_1 + \lambda(Y), Y \in \mathfrak{h}$$

and then the proof of the result is complete. Now from this result we deduce the result that if $\mathfrak{g}$ is a solvable Lie algebra and ρ is a finite dimensional representation of $\mathfrak{g}$ in a complex vector space V, then we can choose a basis B for V such that for each $X \in \mathfrak{g}$, $[\rho(X)]_B$ is an upper triangular matrix. In fact, using Lie's theorem, we first choose a non-zero vector $v_1 \in V$ and a $\lambda \in \mathfrak{g}^*$ so that

$$\rho(X)v_1 = \lambda(X)v_1, X \in \mathfrak{g}$$

Then define

$$V_1 =< v_1 >$$

and observe that since V_1 is ρ-invariant, ρ induces in a natural way a representation ρ_1 of $\mathfrak{g}$ on V/V_1 and again applying Lie's theorem to this representation, we get an element $v_2 \in V, v_2 \notin V_1$ such that $\rho(X)v_2 - \lambda_2(X)v_2 \in V_1, X \in \mathfrak{g}$ for some $\lambda \in \mathfrak{g}^*$. Define

$$V_2 =< v_1, v_2 >$$

Then we have

$$\rho(X)v_2 = \lambda_2(X)v_2 mod V_1, V_1 \subset V_2$$

In general, suppose, we have constructed linearly independent vectors $v_1, ..., v_k$ and linear functionals $\lambda_j \in \mathfrak{g}^*, j = 1, 2, ..., k$ such that

$$\rho(X)v_j = \lambda_j(X)v_j mod < v_1, ..., v_{j-1} >, j = 1, 2, ..., k, X \in \mathfrak{g}$$

Then, the vector space

$$V_k =< v_1, ..., v_k >$$

is clearly ρ-invariant and if $V_k = V$, the proof is complete while if $V_k \neq V$, then ρ induces in a natural way a representation of $\mathfrak{g}$ on the non-zero vector space V/V_k and hence by application of Lie's theorem, we can select a $v_{k+1} \notin V_k$ and a $\lambda_{k+1} \in \mathfrak{g}^*$ such that

$$\rho(X)v_{k+1} - \lambda_{k+1}(X)v_{k+1} = 0 mod V_k$$

Then set,

$$V_{k+1} =< v_1, ..., v_k, v_{k+1} >$$

and the induction proceeds further. This process will terminate in a finite number of steps since V is finite dimensional.

5.3 Engel's theorem on nil-representations of a Lie algebra

Let $\mathfrak{g}$ be a (finite dimensional) Lie algebra and let ρ be a nil-representation of $\mathfrak{g}$ in a finite dimensional complex vector space V, ie, $\rho(X)$ is nilpotent for all $X \in \mathfrak{g}$. Then, there exists a non-zero vector $v \in V$ such that $\rho(X)v = 0 \forall X \in \mathfrak{g}$. Hence, there exists a basis B for V such that $[\rho(X)]_B$ is strictly upper triangular for all $X \in \mathfrak{g}$. In particular, if $\mathfrak{g}$ is a nilpotent Lie algebra, ie, $ad(X)$ is nilpotent for all $X \in \mathfrak{g}$, then, there is a basis B for $\mathfrak{g}$ such that $[ad(X)]_B$ is strictly upper triangular for every $X \in \mathfrak{g}$ and consequently, if $n = dim\mathfrak{g}$ and $X_1, ..., X_n \in \mathfrak{g}$ are arbitrary then

$$ad(X_1).ad(X_2)...ad(X_n) = 0$$

Solution:
Let

$$\mathfrak{a} = Ker(\rho) = \{X \in \mathfrak{g} : \rho(X) = 0\}$$

Then, $\mathfrak{a}$ is an ideal in $\mathfrak{g}$ since if $X \in \mathfrak{g}, Y \in \mathfrak{a}$, then $\rho(Y) = 0$ and therefore,

$$\rho([X, Y]) = [\rho(X), \rho(Y)] = 0$$

implying that $[X, Y] \in \mathfrak{a}$. So it is meaningful to speak of the Lie algebra

$$\mathfrak{g}' = \mathfrak{g}/\mathfrak{a}$$

with the bracket defined by

$$[X + \mathfrak{a}, Y + \mathfrak{a}] = [X, Y] + \mathfrak{a}, X, Y \in \mathfrak{g}$$

Note that the bracket is well defined since

$$X' + \mathfrak{a} = X + \mathfrak{a}, Y' + \mathfrak{a} = Y + \mathfrak{a}$$

imply

$$U = X' - X, V = Y' - Y \in \mathfrak{a}$$

which imply

$$[X', Y'] = [X + U, Y + V] = [X, Y] + Z$$

where

$$Z = [X, V] + [U, Y] + [U, V] \in \mathfrak{a}$$

since $\mathfrak{a}$ is an ideal. Thus,

$$[X', Y'] + \mathfrak{a} = [X, Y] + \mathfrak{a}$$

proving thereby that the bracket on the vector space $\mathfrak{g}/\mathfrak{a}$ is well defined. The necessary properties of the bracket, namely bilinearity, skew-symmetry and the Jacobi identity immediately follows from the same properties of the bracket on $\mathfrak{g}$. Now, clearly the Lie algebra $\mathfrak{g}/\mathfrak{a}$ is isomorphic with the Lie algebra $\rho(\mathfrak{g})$ via the isomorphism ρ_1 that maps $X+\mathfrak{a}$ to $\rho(X)$ for $X \in \mathfrak{g}$. For example, the injectivity of ρ_1 follows from the fact that $X+\mathfrak{a}=\mathfrak{a}$ implies $X \in \mathfrak{a}$ implies $\rho(X)=0$. Surjectivity of ρ_1 is obvious. That ρ_1 preserves the bracket follows from

$$\rho_1([X+\mathfrak{a}, Y+\mathfrak{a}]) = \rho_1([X,Y]+\mathfrak{a})$$

$$= \rho([X,Y]) = [\rho(X), \rho(Y)] = [\rho_1(X+\mathfrak{a}), \rho_1(Y+\mathfrak{a})]$$

Note that ρ_1 is well defined because $X+\mathfrak{a}=X'+\mathfrak{a}$ implies $X'-X \in \mathfrak{a}$ implies $\rho(X')-\rho(X)=\rho(X'-X)=0$. In other words, ρ_1 is a faithful representation of $\mathfrak{g}'$ in V. Since the elements of $\rho(\mathfrak{g})$ are all nilpotent, it follows therefore from the above isomorphism that $\mathfrak{g}'=\mathfrak{g}/\mathfrak{a}$ is a nilpotent Lie algebra, ie, all the elements of $ad(\mathfrak{g}')$ are nilpotent. To see this clearly, we observe that if $U \in \mathfrak{g}'$, then

$$\rho_1 o adU(V) = \rho_1([U,V]) = [\rho_1(U), \rho_1(V)] = ad(\rho_1(U))(\rho_1(V))$$

or equivalently,

$$\rho_1 o ad(U) = ad o \rho_1(U)$$

or equivalently,

$$\rho_1 o ad = ad o \rho_1$$

on $\mathfrak{g}'$ or equivalently,

$$ad|_{\mathfrak{g}'} = \rho_1^{-1} o ad|_{\rho(\mathfrak{g})} o \rho_1$$

on $\mathfrak{g}'$ and since $ad|_{\rho(\mathfrak{g})}$ is nilpotent on $L(V)$ because $\rho(\mathfrak{g})$ is nilpotent on V, it follows that $ad|_{\mathfrak{g}'}$ is also nilpotent on $\mathfrak{g}'$.

Remark: if A is a linear nilpotent operator on a vector space W, then $ad(A)$ is nilpotent on $L(W)$. This follows from the identity

$$ad(A)^n(B) = (L_A - R_A)^n(B) = \sum_{r=0}^{n} \binom{n}{r} L_A^r(-R_A)^{n-r}(B)$$

$$= \sum_{r=0}^{n} \binom{n}{r} (-1)^{n-r} A^r B A^{n-r} \forall B \in L(W)$$

In view of the above remarks, we can assume without any loss of generality that $\mathfrak{g}$ is a nilpotent Lie algebra and ρ is a nil-representation of $\mathfrak{g}$ in order to prove the existence of a non-zero vector v such that $\rho(\mathfrak{g})v=0$.

5.4 Aperture antenna pattern fluctuations

Consider a surface antenna with the surface equation $z=f(x,y)$. Let $\mathbf{E}_i(x,y,z)$ be an electric field at fixed frequency ω that is incident upon this surface.

[a] Justify that the surface magnetic current density on the antenna surface is given by

$$\mathbf{M}_s(x,y) = -\hat{n} \times \mathbf{E}_i(x,y,f(x,y))$$

where

$$\hat{n} = (-f_{,x}(x,y)\hat{x} - f_{,y}(x,y)\hat{y} + \hat{z})/\sqrt{1 + f_{,x}^2 + f_{,y}^2}$$

where

$$f_{,x} = \frac{\partial f}{\partial x}, f_{,y} = \frac{\partial f}{\partial y}$$

[b] Show that the differential surface area element on the antenna surface is given by

$$dS(x,y) = \sqrt{1 + f_{,x}^2 + f_{,y}^2} dxdy$$

[c] Show that the far field electric vector potential radiated by the antenna surface aperture is given by

$$\mathbf{F}(\mathbf{r}) = (\epsilon/4\pi)(exp(-jkr)/r) \int \mathbf{M}_s(x',y')exp(jK\hat{r}.(x'\hat{x}+y'\hat{y}+f(x',y')\hat{z}))dS(x',y')$$

[d] Hence, if the surface fluctuates by a small amount so that its new equation is $z = f(x,y)+\delta f(x,y)$, then evaluate $\delta\mathbf{F}(\mathbf{r})$ and hence $\delta\mathbf{E}(\mathbf{r})$, the radiated fields in the far field zone as a linear functional of δf.

hint:

$$\delta\sqrt{1 + f_{,x}^2 + f_{,y}^2} = (f_{,x}\delta f_{,x} + f_{,y}\delta f_{,y})/\sqrt{1 + f_{,x}^2 + f_{,y}^2}$$

Hence, evaluate $\delta\hat{n}(x,y), \delta dS(x,y)$ in terms of $\delta f(x,y)$ and its partial derivatives.

[6] If $\delta f(x,y)$ in the previous problem is a random function with mean zero and correlations

$$\mathbb{E}(\delta f(x,y).\delta f(x',y'))R_{ff}(x,y|x',y')$$

then evaluate the correlations in the far field pattern fluctuations.

[7] In problem [5], evaluate the total power radiated out by the aperture surface antenna in the far field zone.

hint: In the far field zone, the electric field upto $O(1/r)$ is

$$\mathbf{E} = -\nabla \times \mathbf{F}/\epsilon = jk\hat{r} \times \mathbf{F}/\epsilon$$

and the corresponding far field zone magnetic field is given by

$$-j\omega\mu\mathbf{H} = \nabla \times \mathbf{E} = -jk\hat{r} \times \mathbf{E}$$

Now calculate the far field Poynting vector field $(1/2)Re(\mathbf{E} \times \mathbf{H}^*)$ in the far field zone upto $O(1/r^2)$, take the radial component, multiply it by the surface element $dS = r^2 d\Omega$ where $d\Omega$ is the solid angle differential element and integrate the result over all solid angles to get the total radiated power.

5.5 Spectral theorem using Gelfand-Naimark theorem

[8] Prove the spectral theorem for bounded normal operators on a Hilbert space using the Gelfand-Naimark theorem on spectrum of a commutative Banach algebra combined with the Riesz representation theorem for continuous functionals on the space of continuous functions on a compact metric space.

hint: Let $\mathcal{H}$ be a Hilbert space and A a commutative Banach subalgebra of $B(\mathcal{H})$. Let Δ denote the space of continuous homomorphisms from A into $\mathbb{C}$. Thus, $h \in \Delta$ means that $h : A \to \mathbb{C}$ is a continuous map such that

$$h(TS) = h(T)h(S), T, S \in A, h(I) = 1,$$

$$h(aT + bS) = ah(T) + bh(S), a, b \in \mathbb{C}, T, S \in A$$

Note that if $T \in A$ is invertible, then it follows that

$$h(T^{-1}) = h(T)^{-1}$$

If $T \in A$, then $\lambda \in \sigma(T)$ (the spectrum of T) iff $\lambda - T$ is non-invertible in $A = B(\mathcal{H})$. Note that if $T \in A$, then for $\lambda \in \mathbb{C}$, $\lambda - T \in A$, in particular, this is in $B(\mathcal{H})$ and hence defined on the whole of $\mathcal{H}$, so if it is invertible, then by the open mapping theorem, $(\lambda - T)^{-1}$ is bounded, ie, in A which means that $\lambda \notin \sigma(T)$. Thus, for $T \in A$, we have that $\lambda \in \sigma(T)$ iff $(\lambda - T)$ is non-invertible. If $\lambda \notin \sigma(T)$ and $h \in \Delta$, then

$$h((\lambda - T)^{-1})h(\lambda - T) = 1$$

implies that $h(T) \neq h(\lambda) = \lambda$ ($h(\lambda)$ is an abbreviation for $h(\lambda.e) = \lambda.h(e) = \lambda$). Equivalently, if $h(T) = \lambda$ for some $h \in \Delta$, then $\lambda \in \sigma(T)$ and this implies that $(\lambda - T)$ is not invertible. Note that if we write $\hat{\lambda}(h) = h(\lambda), \lambda \in \mathbb{C}$, then $\hat{\lambda}(h) = \lambda \forall h \in \Delta \forall \lambda \in \mathbb{C}$ and $\hat{T}(h) = h(T), h \in \Delta, T \in A$. Then, it follows that $\lambda \in \sigma(T)$ iff

$$h(T) = \hat{T}(h) = \lambda$$

for some $h \in \Delta$.

Remark: If $\lambda \in \sigma(T)$, then $\lambda - T$ is not invertible and hence $(\lambda - T)^{-1}$ is not defined. Choose a sequence $\lambda_n \to \lambda$ such that $\lambda_n - T$ is invertible for every n and then we get

$$h((\lambda_n - T)^{-1})h(\lambda_n - T) = 1$$

implies

$$h((\lambda_n - T)^{-1})(\lambda_n - h(T)) = 1 \forall n$$

Now since $\lambda - T$ is not invertible and $\lambda_n \to \lambda$, it follows that there must exist an $h \in \Delta$ such that $limsup_n |h((\lambda_n - T)^{-1}| = \infty$ and therefore

$$h(T) = lim \lambda_n = \lambda$$

for such an h.

Remark: Another way to see this is as follows: Suppose $X = \lambda - T$ is not invertible. Then, $I = \{YX : Y \in A\}$ is an ideal in A that does not contain the identity element e and hence I is contained in a maximal proper ideal of A. Thus, there exists an $h \in \Delta$ that vanishes on this ideal and in particular, $h(X) = 0$. From this, it follows that $h(T) = h(\lambda) = \lambda$. In other words, we have proved that

$$\sigma(T) = \{h(T) : h \in \Delta\} = \hat{T}(\Delta)$$

Note that if $h \in \Delta$, then $\lambda\mu = h(\lambda\mu) = h(\lambda)h(\mu), \lambda, \mu \in \mathbb{C}$.

Now let T be a normal operator in $B(\mathcal{H})$, ie, T is bounded and it commutes with T^*. Consider the commutative Banach algebra A generated by $\{T, T^*\}$. Choose $x, y \in \mathcal{H}$ and consider the mapping $\hat{T} \to < x, Ty >$ from $C(\Delta)$ into $\mathbb{C}$. This mapping is well defined since if $T, S \in A$ and $\hat{T} = \hat{S}$, then $h(T) = h(S) \forall h \in \Delta$ and hence $h(T - S) = 0 \forall h \in \Delta$ which implies that $\| T - S \| = 0$, ie, $T = S$.

Remark: if A is a commutative Banach algebra and $x \in A$, then with Δ denoting the space of continuous homomorphisms on A, we have that

$$\| x \| = sup\{h(x) : h \in \Delta\} = sup\{\hat{x}(h) : h \in \Delta\} = \| \hat{x} \|$$

Remark: We have seen that the spectrum $\sigma(T)$ of T can be identified with the continuous function $\hat{T}$ on Δ. In fact, we have seen that if A is a commutative Banach algebra, then

$$\sigma(T) = \hat{T}(\Delta) = \{h(T) : h \in \Delta\}$$

For example, if T is a bounded Hermitian operator in the Hilbert space $\mathfrak{H}$ and if A is the Banach algebra of all bounded functions of T, then from the spectral representation of T, then for any bounded function f on $\mathbb{R}$, we have

$$f(T) = \int f(\lambda)dE(\lambda) = \int \lambda dE(f^{-1}(-\infty, \lambda])) - -(1)$$

We note that if $S_1, S_2 \in A$,

$$S_k = f_k(T) = \int f_k(\lambda)dE(\lambda), k = 1, 2$$

for some bounded measurable functions $f_k, k = 1, 2$ on $\mathbb{R}$ and therefore, if $h \in \Delta$ where Δ is the space of continuous homomorphisms from A into $\mathbb{C}$, then

$$h(S_1S_2) = \int f_1(\lambda)f_2(\lambda)h(dE(\lambda)) =$$

$$\int f_1(\lambda)f_2(\lambda)dh(E(\lambda))$$

on the one hand and on the other,

$$h(S_1S_2) = h(S_1)h(S_2) = \int f_1(\lambda)f_2(\mu)h(dE(\lambda)).h(dE(\mu))$$

$$= \int f_1(\lambda)f_2(\mu)dh(E(\lambda))dh(E(\mu))$$

and comparing these two expressions, we can infer that

$$dh(E(\lambda)).dh(E(\mu)) = dh(E(\lambda))\delta_{\lambda,\mu}$$

which is equivalent to saying that

$$h(E(B_1)).h(E(B_2)) = h(E(B_1 \cap B_2)), B_1, B_2 \in \mathcal{B}(\mathbb{R})$$

This is in agreement with the homomorphism property of $h : A \to \mathbb{C}$:

$$h(E(B_1 \cap B_2)) = h(E(B_1).E(B_2)) = h(E(B_1)).h(E(B_2)), B_1, B_2 \in \mathcal{B}(\mathbb{R})$$

Now, suppose, we pick a $\lambda \in \sigma(T)$ and define

$$h_\lambda(T) = \lambda,$$

and more generally, for

$$S = \int f(\lambda)dE(\lambda)$$

we define

$$h_\lambda(S) = f(\lambda)$$

then

$$h_\lambda(S_1S_2) = f_1(\lambda)f_2(\lambda) = h_\lambda(S_1)h_\lambda(S_2),$$

$$h_\lambda(c_1S_1 + c_2S_2) = c_1f_1(\lambda) + c_2f_2(\lambda) =$$

$$c_1h_\lambda(S_1) + c_2h_\lambda(S_2)$$

and hence $h_\lambda : A \to \mathbb{R}$ is a homomorphism, ie,

$$h_\lambda \in \Delta$$

It is clear from the above formulas that

$$sup\{|h_\lambda(S)| : \lambda \in \sigma(T)\} = sup\{|f(\lambda)| : \lambda \in \sigma(T)\} = \| S \|$$

More generally, if $h \in \Delta$ is arbitrary, ie, $h : A \to \mathbb{C}$ is a homomorphism, then for

$$S = \int f(\lambda)dE(\lambda)$$

we find that

$$h(S) = \int f(\lambda)dh(E(\lambda))$$

where the integral is taken over $\lambda \in \sigma(T)$. We now observe that for any Borel function g on $\mathbb{R}$, we have that

$$h(g(T)) = g(h(T)), h \in \Delta$$

where $g(T)$ is defined as an operator via the spectral theorem for this result follows from

$$h(T^n) = h(T)^n, n = 0, 1, ...$$

and hence by taking linear combinations,

$$h(g(T)) = g(h(T))$$

In particular, taking g as the function which maps T to $E(\lambda)$, ie, $g(x) = \chi_{(-\infty,\lambda]}(x)$, we get

$$h(E(\lambda)) = \chi_{(-\infty,\lambda]}(h(T))$$

and hence,

$$h(S) = \int f(\lambda)d_\lambda \chi_{(-\infty,\lambda]}(h(T))$$

$$= \int f(\lambda)d\theta(\lambda - h(T)) = f(h(T))$$

which is consistent with our above observation. Thus,

$$\| S \| = sup\{|f(\lambda)| : \lambda \in \sigma(T)\}$$

$$= sup\{|f(h(T))| : h \in \Delta\}$$

which is once again, consistent with our previous observations regarding the definition of the spectrum of T as the set of all λ for which $\lambda - T$ is not invertible and our result that this spectrum is also equal to the set of all $h(T)$ as h varies over Δ.

Now we return to the commutative Banach algebra A generated by T, T^* where T is a bounded normal operator. Then, we get from the Riesz representation theorem, in view of the fact that $\hat{S} \rightarrow < x, Sy >$ from $C(\Delta)$ into $\mathbb{C}$ is a bounded linear functional (in fact,

$$\| S \| = sup\{|h(S)| : h \in \Delta\} =$$

$$= sup\{|\hat{S}(h)| : h \in \Delta\} = \| \hat{S} \|$$

where the last norm occurs because it is the standard *sup*-norm on the space $C(\Delta)$)

$$< x, Sy > = \int_\Delta \hat{S}(h)d\mu_{x,y}(h), S \in A$$

for some measure $\mu_{x,y}$ on Δ. Replacing S by $f(T)$ gives us

$$< x, f(T)y >= \int_{\Delta} \Phi(f)(h) d\mu_{x,y}(h)$$

where we have defined

$$\Phi(f) = \hat{S} \in C(\Delta), S = f(T)$$

where f is a bounded Borel function on $\mathbb{R}$. Now writing $S_1 = f_1(T), S_2 = f_2(T)$, we get

$$\Phi(f_1 f_2)(h) = (S_1 S_2)\hat{}(h) = h(S_1 S_2) = h(S_1)h(S_2)$$

$$= \hat{S}_1(h)\hat{S}_2(h) = \Phi(f_1)(h)\Phi(f_2)(h) = (\Phi(f_1)\Phi(f_2))(h), h \in \Delta$$

and hence

$$\Phi(f_1 f_2) = \Phi(f_1)\Phi(f_2)$$

for all bounded Borel functions f_1, f_2 on $\mathbb{R}$. In particular, if B_1, B_2 are bounded Borel subsets of $\mathbb{R}$, then we get

$$\Phi(\chi_{B_1 \cap B_2}) = \Phi(\chi_{B_1}).\Phi(\chi_{B_2})$$

We also have the following useful identities:

$$\int \Phi(f_1)(h) d\mu_{f_2(T)x,y}(h) =< f_2(T)x, f_1(T)y >=< x, f_2(T)^* f_1(T)y >$$

$$=< f_1(T)^* x, f_2(T)^* y >= \mu_{f_1(T)^* x, f_2(T)^* y}(\Delta)$$

In fact, suppose $S \in A$, ie, S is a Borel function of T, T^*. Then,

$$< x, STy >=< S^* x, Ty >=< x, TSy >=< T^* x, Sy >$$

and hence,

$$\mu_{x,STy}(\Delta) = \mu_{S^* x, Ty}(\Delta) = \mu_{x,TSy}(\Delta) = \mu_{T^* x, Sy}(\Delta)$$

For

$$S = f(T, T^*)$$

we define $\Psi(f) = \hat{S} \in C(\Delta)$ and then get

$$< x, Sy >= \int_{\Delta} h(S) d\mu_{x,y}(h)$$

Replacing S by $\chi_B(S)$ where B is a bounded Borel set in $\mathbb{R}$ gives us

$$< x, \chi_B(S)y >= \int_{\Delta} h(\chi_B(S)) d\mu_{x,y}(h)$$

and thus, if B_1, B_2 are two bounded Borel subsets of $\mathbb{R}$, we have

$$< x, \chi_{B_1 \cap B_2}(S)y >= \int_\Delta h(\chi_{B_1}(S))h(\chi_{B_2}(S))d\mu_{x,y}(h)$$

Remark: To relate this circle of ideas to the spectral theorem for normal operators, we must transform the integral over Δ to an integral over $\sigma(T) = \hat{T}(\Delta)$ Since

$$\hat{T}(\Delta) = \{h(T) : h \in \Delta\} = \sigma(T),$$

it follows that

$$< x, Ty >= \int_\Delta \hat{T}(h)d\mu_{x,y}(h) = \int_{\sigma(T)} \lambda d\mu_{x,y}o\hat{T}^{-1}(\lambda)$$

and more generally,

$$< x, f(T)y >= \int_\Delta \Phi(f)(h)d\mu_{x,y}(h)$$

with

$$\Phi(f_1 f_2) = \Phi(f_1)\Phi(f_2), \Phi(cf_1 + f_2) = c\Phi(f_1) + \Phi(f_2)$$

More importantly,

$$< x, f(T)y >= \int_\Delta h(f(T))d\mu_{x,y}(h)$$

Now if $f(\lambda) = \lambda^n$, then

$$h(f(T)) = h(T^n) = h(T)^n = f(h(T))$$

and since h is linear, it follows that if f is any polynomial. then

$$h(f(T)) = f(h(T))$$

and by taking limits and using the continuity of h we get that this identity is true even if f is any bounded Borel function. Thus,

$$< x, f(T)y >= \int_\Delta f(h(T))d\mu_{x,y}(h)$$

$$= \int_\Delta fo\hat{T}(h)d\mu_{x,y}(h) = \int_{\sigma(T)} f(\lambda)d\mu_{x,y}o\hat{T}^{-1}(\lambda)$$

which is precisely the content of the spectral theorem once we are able to show that for any bounded Borel set B in $\mathbb{R}$, we can write

$$\mu_{x,y}(\hat{T}^{-1}(B)) =< x, E_T(B)y >$$

where E_T is a spectral measure on the Borel subsets of $\mathbb{R}$ with values in the space of orthogonal projection operators on $\mathcal{H}$.

5.6 The Atiyah-Singer index theorem:A supersymmetric proof

Prerequisites:

[1] The Dirac operator in curved space-time with a Yang-Mills connection term.

[2] Lichnerowicz' formula for the square of the Dirac operator in curved space time with a Yang-Mills connection term.

[3] The index of a linear operator.

[4] Expressing the index of a linear operator D using the difference trace of the heat kernels $exp(-tD^*D)$ and $exp(-tDD^*)$ by observing that the non-zero eigenvalues of D^*D and DD^* are identical inclusive of multiplities and hence the contribution to the index comes only from the multiplicities of their zero eigenvalues.

[5] Write

$$T = \begin{pmatrix} 0 & D^* \\ D & 0 \end{pmatrix}$$

and observe that

$$T^2 = \begin{pmatrix} D^*D & 0 \\ 0 & DD^* \end{pmatrix}$$

and hence with str denoting supertrace, we have

$$str(exp(-tT^2)) = Tr(exp(-tD^*D)) - Tr(exp(-tDD^*)) = Index(D), t \in \mathbb{R}$$

5.7 Replicas, regular elements, Jordan decomposition and Cartan subalgebras

[6] Describe Chevalley's theory of replicas and how it is used in proving the Jordan decomposition of a semisimple Lie algebra $\mathfrak{g}$. The steps involved are as follows:

[a] Choose any $X \in \mathfrak{g}$ and consider the derivation

$$D = ad(X)$$

of $\mathfrak{g}$.

[b] Write down the Jordan decomposition of D viewed as a linear operator on $\mathfrak{g}$:

$$D = U + V$$

where U, V are respectively the semisimple and nilpotent parts of D. Thus $[U, V] = 0$ and hence $[X, U] = [X, V] = 0$ and in fact, we know from basic linear algebra that U and V are expressible as polynomial functions of D having constant coefficient zero and that U, V with this constaint on semisimplicity and nilpotency and mutual commutativity are uniquely determined by D.

[c] Hence, it is obvious that

$$D_{r,s} = U_{r,s} + V_{r,s}, r, s \geq 0, r + s \geq 1$$

is the Jordan decomposition of $D_{r,s}$ on $W_{r,s} = W^{\otimes r} \otimes W^{*s}$ with $V = \mathfrak{g}$.

[d] Thus $U_{r,s}, V_{r,s}$ are expressible as polynomial functions of $D_{r,s}$ with these polynomials having zero constant coefficient. Hence

$$N(D_{r,s}) \subset N(U_{r,s}), N(V_{r,s}), \forall r, s$$

ie, U, V are replicas of D. In particular, taking $r = 1, s = 2$ gives us $N(D_{1,2}) \subset N(U_{1,2}), N(V_{1,2})$ which means that U, V are also derivations on $V = \mathfrak{g}$. Since every derivation of a semisimple Lie algebra is inner (this is proved using the non-singularity of the Cartan-Killing form), it follows that $U = ad(S), V = ad(N)$ where $S, N \in \mathfrak{g}$. Since U and V commute and $ad([S, N]) = [ad(S), ad(N)]$, it follows from the faithfulness of the ad map on $\mathfrak{g}$ that $[S, N] = 0$ and hence, we obtain the Jordan decomposition on the semisimple Lie algebra $\mathfrak{g}$:

$$X = S + N, S, N \in \mathfrak{g},$$

$ad(S)$ is semisimple, $ad(N)$ is nilpotent and $[S, N] = 0$. Furthermore, this decomposition is unique as follows from the Jordan decomposition on a vector space and the faithfulness of the ad map on semisimple Lie algebras.

[7] Show that if $\mathfrak{g}$ is a semisimple Lie algebra, and $\mathfrak{h}$ is a Cartan subalgebra, then there exists a regular element $X \in \mathfrak{g}$ such that

$$\mathfrak{h} = \mathfrak{h}_X = \bigcup_{m \geq 1} N(ad(X)^m)$$

hint: Let $X \in \mathfrak{h}$ be a regular element. Since $ad(X)$ is nilpotent on $\mathfrak{h}$ and since X is regular, it follows that $ad(X)$ is non-singular on $\mathfrak{g}/\mathfrak{h}$. Then, it is easy to see that $\mathfrak{h}_X \subset \mathfrak{h}$ and by regularity of X, we get that $\mathfrak{h}_X = \mathfrak{h}$.

[8] Show that if $\mathfrak{g}$ is a semisimple Lie algebra, and $\mathfrak{h}$ is a Cartan subalgebra, then $\mathfrak{h}$ is maximal Abelian. hint:The steps involved in the proof ar outlined below:

[a] Let X be regular in $\mathfrak{g}$ such that $\mathfrak{h} = \mathfrak{h}_X$. Let

$$X = S + N$$

be its Jordan decomposition. Then since $ad(X)(S) = [X, S] = 0, ad(X)(N) = [X, N] = 0$, it follows that $S, N \in \mathfrak{h}_X = \mathfrak{h}$. It is also clear that S is regular and hence $\langle_S = \langle$. Regularity of S can be seen as follows. Clearly, since $ad(S)$ and $ad(N)$ commute, $ad(N)$ leaves each eigensubspace of $ad(S)$ invariant and on such a subspace, $ad(N)$ is nilpotent. Thus, there is a basis for $\mathfrak{g}$ relative to which $ad(S)$ is diagonal and $ad(N)$ is strictly upper triangular. It follows that relative to such a basis $ad(X)$ is upper triangular with the same diagonal entries as those of $ad(S)$. Hence, the characteristic polynomial of $ad(X)$ is the same as that of $ad(S)$ and since X is regular, it follows that S is also regular. Now we wish to show that $N = 0$ and that will prove that $ad(\mathfrak{h})$ consists of only semisimple elements.

Another proof of the fact that $\mathfrak{h}_S = \mathfrak{h}_X$. First observe that $ad(X)(S) = 0$ and hence $S \in \mathfrak{h}_X$. Suppose that $Y \in \mathfrak{h}_S$. Then, $ad(S)^m(Y) = 0$ for some positive m and since $ad(S)$ is semisimple, it follows that $ad(S)(Y) = 0$. It follows that $ad(S)^m(Y) = 0$ for all $m > 0$ and therefore,

$$ad(X)^m(Y) = \sum_{r=0}^{m} \binom{m}{r} ad(N)^{m-r} ad(S)^r(Y) = ad(N)^m(Y) = 0$$

for sufficiently large m since $ad(N)$ is nilpotent. This implies that

$$Y \in \mathfrak{h}_X$$

and therefore,

$$\mathfrak{h}_S \subset \mathfrak{h}_X$$

Conversely suppose $Y \in \mathfrak{h}_X$. Then,

$$ad(S)^m(Y) = (ad(X) - ad(N))^m(Y) = \sum_{r=0}^{m} (-1)^{m-r}(ad(X))^r(ad(N))^{m-r}(Y)$$

for sufficiently large m since $ad(X)$ is nilpotent on $\mathfrak{h}_X$ while $ad(N)$ is nilpotent. This proves that

$$Y \in \mathfrak{h}_S$$

and hence,

$$\mathfrak{h}_X = \mathfrak{h}_S$$

Another way to see that $\mathfrak{h}_X = \mathfrak{h}_S$ is simply to prove that S is a regular element of $\mathfrak{h}_X$. But this would follow immediately if we are able to prove that the characteristic polynomial of $ad(S)$ equals that of $ad(X)$. This can be seen in the following way:

$$ad(X) = ad(S) + ad(N), [ad(S), ad(N)] = 0$$

is the Jordan decomposition of $ad(X)$. Now $ad(S)$ is semisimple on $\mathfrak{g}$ and hence we can choose a basis B for $\mathfrak{g}$ such that $[ad(S)]_B$ is diagonal. Let $c_1, .., c_r$ be the distinct eigenvalues of $ad(S)$, then we can write the spectral decomposition of $ad(S)$ as

$$ad(S) = \sum_{k=1}^{r} c_k E_k$$

where

$$I = \sum_{k=1}^{r} E_k$$

is a resolution on I (not necessarily orthogonal). Since $ad(N)$ commutes with $ad(S)$, it follows that it also commutes with every E_k, ie, it leaves $R(E_k) = N(ad(N) - c_k I)$ invariant. Now we can choose a basis for $R(E_k)$ so that $ad(N)$ restricted to $R(E_k)$ is strictly upper triangular since $ad(N)$ restricted to E_k is nilpotent. If we pool up all these bases for $R(E_k)$ then we get a basis for $\mathfrak{g}$ such that $ad(S)$ in this basis is diagonal and $ad(N)$ in this basis is strictly upper triangular. In other words, we have shown that $ad(X) = ad(S) + ad(N)$ in this basis is upper-triangular with same diagonal entries as that of $ad(S)$. This immediately proves that the characteristic polynomials of $ad(X)$ and $ad(S)$ are the same and therefore since X is regular, it follows that S is also regular.

Problem: Show that if X is regular then $Y \in \mathfrak{h}_X$
$= \mathfrak{h}$ is regular iff $det(ad(Y)|_{\mathfrak{g}/\mathfrak{h}}) \neq 0$

hint: The above determinant is zero iff there exists a $Z_1 \in \mathfrak{g} - \mathfrak{h}$ such that

$$ad(Y)(Z_1) = 0$$

Now choose elements $Z_2, ..., Z_r$ such that $\{Z_k + \mathfrak{h} : k = 1, 2, ..., r\}$ form a basis for $\mathfrak{g}/\mathfrak{h}$ and show that

$$\mathfrak{g} = \mathfrak{h} \oplus \mathfrak{q}$$

where

$$\mathfrak{q} = span\{Z_1, ..., Z_m\}$$

Now choose a basis $\{H_1, ..., H_l\}$ for $\mathfrak{h}$ such that $ad(Y)|_{\mathfrak{h}}$ in this basis is strictly upper triangular (This is possible because $\mathfrak{h}$ is a nilpotent Lie algebra). Then compute the characteristic polynomial of $ad(Y)$ in the basis $\{H_1, ..., H_l, Z_1, ..., Z_r\}$ for $\mathfrak{g}$ and observe that if $ad(Y)(Z_1) = 0$, then this characteristic polynomial is of the form

$$t^{l+1} f(t)$$

where f is a polynomial and hence Y cannot be regular. Further, if there is no such vector Z_1, then show that the characteristic polynomial of $ad(Y)$ relative to this basis must be of the form $t^l f(t)$ where f is a polynomial with $f(0) \neq 0$ or in other words, Y is regular. To show this, we must make the fact that $\mathfrak{h}_X$ is a nilpotent Lie algebra for regular X. To prove this fact, let $Y \in \mathfrak{h}_X$ be such that the determinant of $ad(Y)$ on $\mathfrak{g}/\mathfrak{h}_X$ is non-zero. Then, it is clear that the characteristic polynomial of $ad(Y)$ must be of the form $t^k f(t)$ where $k \leq dim\mathfrak{h}_X$ and $f(0) \neq 0$. Hence, $\mathfrak{h}_Y \subset \mathfrak{h}_X$. This is possible only if $\mathfrak{h}_Y = \mathfrak{h}_X$ because X is regular and is nilpotent on $\mathfrak{h}_X$ and non-singular on $\mathfrak{g}/\mathfrak{h}_X$ and further that $\mathfrak{h}_X$ is $ad(Y)$-invariant (since $ad(X)^m(Y) = 0$ for some positive m, it follows that $ad(ad(X))^m(ad(Y)) = 0$ and therefore from elementary linear algebra, $ad(Y)$ leaves $\mathfrak{h}_X = \bigcup_{m \geq 1} N(ad(X)^m)$ invariant). In particular, $ad(Y)$ is nilpotent on $\mathfrak{h}_X$. Since the elements Y of $\mathfrak{h}_X$ for which $ad(Y)$ is non-singular on $\mathfrak{g}/\mathfrak{h}_X$ are dense in $\mathfrak{h}_X$, it follows that every $Y \in \mathfrak{h}_X$ is such that $ad(Y)$ is nilpotent on $\mathfrak{h}_X$. Note that this proof also shows that if Y is a regular element of $\mathfrak{h}_X$, then $\mathfrak{h}_Y = \mathfrak{h}_X$.

Now coming back to the proof that every Cartan algebra $\mathfrak{h}$ is maximal abelian, when $\mathfrak{g}$ is semisimple we choose a regular X in $\mathfrak{h}$ and observe that $\mathfrak{h} = \mathfrak{h}_X$ because, $ad(X)$ is nilpotent on $\mathfrak{h}$ (by that part in the definition of a Cartan algebra which states that a Cartan algebra is nilpotent) and hence $\mathfrak{h} \subset \mathfrak{h}_X$. Further, consider the Lie algebra $\mathfrak{h}_X/\mathfrak{h}$. By Engel's theorem applied to the adjoint representation of the nilpotent algebra $\mathfrak{h}$ on this space, we deduce that there exists a $Y \in \mathfrak{h}_X - \mathfrak{h}$ such that $[Y, \mathfrak{h}] \subset \mathfrak{h}$ and since by that part in the definition of a Cartan algebra which states that it is its own normalizer, it follows that $Y \in \mathfrak{h}$, a contradiction unless $\mathfrak{h} = \mathfrak{h}_X$. (This argument does not require $\mathfrak{g}$ to be semisimple Now, for semisimple $\mathfrak{g}$, we let

$$X = S + N$$

be the Jordan decomposition of X where X is a regular element in $\mathfrak{h}$. We wish to show that $N = 0$. (Note that the existence of the Jordan decomposition depends on the semisimplicity of $\mathfrak{g}$. Since $ad(S)$ is semisimple, we have the decomposition

$$\mathfrak{g} = R(ad(S)) \oplus N(ad(S))$$

Since S is a regular element of $\mathfrak{h} = \mathfrak{h}_X$, it follows that $\mathfrak{h}_S = \mathfrak{h}$ and since $ad(S)$ is semisimple, it follows that

$$\mathfrak{h} = \mathfrak{h}_S = N(ad(S))$$

So we can write

$$\mathfrak{g} = R(ad(S)) \oplus \mathfrak{h}$$

Now since $[S, N] = 0$, it follows that for any $Y \in \mathfrak{g}$, we have

$$< N, ad(S)(Y) >=< N, [S, Y] >=< [N, S], Y >= 0$$

and hence by the nonsingularity of $< ., . >$ for semisimple Lie algebras, it follows that

$$N \perp R(ad(S))$$

To prove that $N \perp \mathfrak{h}$, we take any $Y \in \mathfrak{h} = N(ad(S))$. We must show that

$$Tr(ad(N).ad(Y)) =< N, Y >= 0$$

Recall that $ad(X)(S) = ad(S)(N) = 0$ and hence $S, N \in \mathfrak{h}_X = \mathfrak{h}$. Now every nilpotent Lie algebra is also solvable. (This can be seen by making use of Engel's theorem for nil-representations. Indeed let $\mathfrak{g}_0$ be a nilpotent Lie algebra. Then the adjoint representation of $\mathfrak{g}_0$ on itself is a nil representation and hence according to Engel's theorem, there exists a basis for $\mathfrak{g}_0$ relative to which all the operators $ad(Z), Z \in \mathfrak{g}_0$ are strictly upper triangular. Hence, by working in this basis, we deduce that for all sufficiently large n, we have

$$ad(Z_1).ad(Z_2)..ad(Z_n) = 0 \forall Z_1, ..., Z_n \in \mathfrak{g}_0$$

which is the same as saying that

$$[Z_1,[Z_2,..,[Z_{n-1},[Z_n,Z]]...,]=0, forall Z_1,...,Z_n,Z \in \mathfrak{g}_0$$

and in particular,

$$D^n\mathfrak{g}_0=0$$

thereby establishing solvability of $\mathfrak{g}_0$). Thus, $\mathfrak{h}$ is a solvable Lie algebra and hence by applying Lie's theorem to the adjoint representation of $\mathfrak{h}$ in $\mathfrak{g}$, we deduce that there exists a basis for $\mathfrak{g}$ relative to which all the operators $ad(H), H \in \mathfrak{h}$ are upper-triangular matrices. In particular, relative to this basis $ad(N)$ and $ad(Y)$ are also upper-triangular. But since $ad(N)$ is nilpotent on $\mathfrak{g}$, it must necessarily follow that relative to this basis, $ad(N)$ is strictly upper-triangular. Since the product of an upper-triangular matrix and a strictly upper-triangular matrix is strictly upper-triangular, it follows that $ad(N).ad(Y)$ is strictly upper-triangular and hence its trace is zero, ie,

$$<N,Y>=0$$

Therefore, we have proved that $N \perp R(ad(S)) \oplus N(ad(S)) = \mathfrak{g}$ and therefore, $N=0$ by non-degeneracy of the Cartan-Killing symmetric bilinear form on $\mathfrak{g}$. Hence $X=S$. In other words, we have proved that if X is a regular element of $\mathfrak{h}$, then X is semisimple (ie, $ad(X)$ is semisimple on $\mathfrak{g}$). Now for such an X, we have that since $\mathfrak{h}=\mathfrak{h}_X$ the result that $ad(X)^m(\mathfrak{h})=0$ for sufficiently large m and therefore (by expressing the semisimple operator $ad(X)$ on $\mathfrak{h}$ relative to basis which makes it diagonal) that $ad(X)(\mathfrak{h})=0$, ie, $[X,\mathfrak{h}]=0$. Now if $Y \in \mathfrak{h}$ is arbitrary, we can write $Y=limX_n$ where $X_n \in \mathfrak{h}$ are semisimple because the set of regular elements of $\mathfrak{h}$ is dense in $\mathfrak{h}$ and we have shown that every regular element of $\mathfrak{h}$ is semisimple in $\mathfrak{g}$. Hence,

$$[Y,\mathfrak{h}]=lim[X_n,\mathfrak{h}]=0$$

proving that $\mathfrak{h}$ is Abelian. Another way to see this is to start with the equation

$$[X,\mathfrak{h}]=0\forall X \in \mathfrak{h}'$$

and hence,

$$[\mathfrak{h}',\mathfrak{h}']=0$$

with $ad(\mathfrak{h}')$ consisting only of semisimple elements. Thus, $ad(\mathfrak{h}')$ is simultaneously diagonable and hence there is a basis B for $\mathfrak{g}$ so that $[ad(\mathfrak{H})]_B$ is diagonal for all $H \in \mathfrak{h}'$. Then taking the limit points of this set we get that $[ad(H)]_B$ is diagonal for all $H \in \mathfrak{h}$.

Remark: We are making use of the fact that the space of regular elements of any Lie algebra $\mathfrak{g}$ is dense in $\mathfrak{g}$. To prove this result, we choose an element $Y \in \mathfrak{g}$ that is non-regular, ie, if $l=rk(\mathfrak{g})$ (then,

$$det(tI-ad(Y))=c(l+1)t^{l+1}+...-t^n$$

Let $Z \in \mathfrak{g}$ be arbitrary and consider

$$f(t,s) = det(t - ad(X + sZ)) = det(tI - ad(X) - s.ad(Z))$$

This is a polynomial in t, s and hence it can be expressed as

$$f(t,s) = c_l(s)t^l + c_{l+1}(s)t^{l+1} + ... + c_n(s)t^n$$

where $c_l(s)$ is a polynomial in s. Suppose that $c_l(s)$ vanishes for all $|s| < \delta$. Then it vanishes for all s since $c_l(s)$ being a polynomial has only a finite number of zeroes. In this case, it would follow that $X + sZ$ is an irregular element for all s. By choosing a basis $\{Z_1, ..., Z_n\}$ for $\mathfrak{g}$ and applying this argument, we would deduce that any element of the form $X + s_1Z_1 + ... + s_nZ_n$ is irregular and hence that there exists no regular element in $\mathfrak{g}$ which is a contradiction to the definition of $l = rk(\mathfrak{g})$. This means that there exists an element $Z \in \mathfrak{g}$ such that for each $\delta > 0$ there is an s with $|s| < \delta$ such that $X + sZ$ is regular. In other words, there is a sequence $s_n \to 0$ such that $X + s_nZ$ is regular for all $n = 1, 2, ...$ Writing $X_n = X + s_nZ$, we deduce that X_n is regular for each n and $limX_n = X$. The proof is complete.

Remark: If $\mathfrak{g}$ is any Lie algebra and $\mathfrak{h}$ is a CSA (ie, $\mathfrak{h}$ is a nilpotent Lie subalgebra of $\mathfrak{g}$ and is also its own normalizer in $\mathfrak{g}$, then $\mathfrak{h}$ is also maximally nilpotent, ie, every CSA is maximally nilpotent. Note that when we say that $\mathfrak{h}$ is a nilpotent sublgebra, we mean that $[\mathfrak{h}, \mathfrak{h}] \subset \mathfrak{h}$ and secondly that $ad(H)^m(\mathfrak{h}) = 0 \forall H \in \mathfrak{h}$ where m some finite positive integer. Note that since we are dealing only with finite dimensional Lie algebras, nilpotency of $\mathfrak{h}$ is equivalent to saying that for each $H_1 \in \mathfrak{h}$ there is a finite positive integer m such that $ad(H_1)^m(H_2) = 0 \ \forall H_2 \in \mathfrak{h}$.

To see the maximal nilpotency of a CSA $\mathfrak{h}$, suppose $\mathfrak{h} \subset \mathfrak{h}_1$ is a proper inclusion where $\mathfrak{h}_1$ is a nilpotent subalgebra. Then since $\mathfrak{h}$ is $ad(\mathfrak{h})$-invariant, it follows that $ad : \mathfrak{h} \to L(\mathfrak{h}_1/\mathfrak{h})$ is a well defined representation and since $ad(\mathfrak{h}_1)$ is nilpotent on $\mathfrak{h}_1$ and $\mathfrak{h} \subset \mathfrak{h}_1$, it follows that $ad(\mathfrak{h})$ is nilpotent on $\mathfrak{h}_1/\mathfrak{h}$. In other words, this representation is a nil-representation and hence by Engel's theorem, there exists a $Y \in \mathfrak{h}_1 - \mathfrak{h}$ such that $ad(\mathfrak{h})(Y + \mathfrak{h}) = \mathfrak{h}$ which is equivalent to saying that $[Y, \mathfrak{h}] \subset \mathfrak{h}$ and since $\mathfrak{h}$ is its own normalizer in $\mathfrak{g}$, it follows that $Y \in \mathfrak{h}$, a contradiction.

5.8 Lecture Plan, Matrix Theory

[1] Vector spaces, basis, linear transformations

[2] Matrix representation of a linear transformation relative to a basis, similarity transformations.

[3] Rank and nullity of a matrix.

[4] Decomposition theorems for matrices.

[a] QR factorization based on the Gram-Schmidt orthonormalization process.

[b] LDU decomposition of a positive definite matrix.

[e] Spectral theorem for normal operators in finite and infinite dimensional Hilbert spaces.

[f] Polar decomposition in a Hilbert space.

[g] Singular value decomposition in finite dimensional Hilbert spaces.

[5] Primary decomposition theorem of a matrix in a finite dimensional complex vector space.

[6] Jordan decomposition of a matrix.

[7] Canonical representation of nilpotent matrices.

[8] The Jordan canonical form.

[9] Computational algorithms for calculating the Jordan canonical form.

[10]Perturbation theory for computing the eigenvalues and eigenvectors of Hermitian and diagonable matrices.

[11] Calculating functions of matrices using contour integrals in the complex plane.

[12] Computing functions of matrices using the Jordan canonical form.

[13] Matrix norms and Banach algebras.

[14] Homomorphisms and spectra of commutative Banach algebras.

[15] Numerical methods for computing eigenvalues and eigenvectors of a matrix.

[16] Tensor product of vector spaces and of linear transformations.

[17] Application of the tensor product to describing nonlinear systems.

[18] Quotient vector spaces and linear transformations on them.

[19] Lie groups, Lie algebras and their representations.

[19.1] Definition of a Lie group and its Lie algebra.

[19.2] Jacobi identity on a Lie algebra.

[19.3] Representation of a Lie group and the associated representation of its Lie algebra.

[19.4] The adjoint representation of a Lie algebra.

[19.5] Solvable and semisimple Lie algebras.

[19.6] Nilpotent Lie algebras and nilpotent representations of a Lie algebra.

[19.7] Lie's theorem on reprsentations of solvable Lie algebras.

[19.8] Engel's theorem on nilpotent representations of a Lie algebra.

[19.9] Regular elements of a Lie algebra.

[19.10] Cartan subalgebras of a Lie algebra and their construction.

[19.11] Properties of Cartan subalgebras.

[19.12] Jordan decomposition on a Lie algebra, proofs based on Chevalley's theory of replicas.

[19.13] Cartan subalgebras of a semisimple Lie algebra and their properties.

[19.14] The root space decomposition of a semisimple Lie algebra.

[19.15] Cartan's classification of all the complex simple Lie algebras.

[20] Some notions in the theory of unbounded operators in a Hilbert space.
[20.1] The uniform boundedness principle.
[20.2] The Hahn-Banach theorem.
[20.3] The closed graph theorem.
[20.4] Closed operators.
[20.5] Adjoint of an operator.
[21] Applications of unbounded operators to quantum mechanics.

Test on Probability theory

[1] Let $(\Omega, \mathcal{F}, P)$ be a probability space and let $E_n, n = 1, 2, ...$ be an infinite sequence of events on this space, ie $E_n \in \mathcal{F}, n = 1, 2, ...$ Then justify the statement that the event that an infinite number of the E_n's occur is given by

$$\{E_n, i.o\} = \bigcap_{n\geq 1}\bigcup_{k\geq n} E_k$$

Show further that the probability of this event satisfies

$$P(\{E_n, i.o\}) = lim_{n\to\infty}\sum_{n\geq 1} P(\bigcup_{k\geq n} E_k)$$

$$\leq \sum_{k\geq n} P(E_k)$$

and in particular, show that if

$$\sum_{k\geq 1} P(E_k) < \infty$$

then the probability of an infinite number of E_n's occurring is zero.

[2] Show that if $X_n, n \geq 1$ is an infinite sequence of random variables on the same probability space, then X_n converges to zero with probability one if for each $\epsilon > 0$,

$$\sum_{n\geq 1} P(|X_n| > \epsilon) < \infty$$

hint: Show that the event that X_n does not converges to zero can be expressed as

$$\{X_n \to 0\}^c = \bigcup_{k\geq 1}\{|X_n| > 1/k, i.o\}$$

and that this event has probability zero if

$$P(\{|X_n| > 1/k, i.o\}) = 0, k = 1, 2, ..$$

Now make use the result of the preceding problem.

[3] Let $(\Omega, \mathcal{F}, P)$ be a probability space and let ϕ be a continuous convex bounded function. By convex, we mean that

$$\phi(\lambda.x + (1-\lambda)y) \leq \lambda.\phi(x) + (1-\lambda)\phi(y)\forall x, y \in \mathbb{R}, 0 \leq \lambda \leq 1$$

Let $X \in L^1(\Omega, \mathcal{F}, P)$, ie, $\mathbb{E}|X| < \infty$. Then prove Jensen's inequality:

$$\mathbb{E}(\phi(X)) \geq \phi(\mathbb{E}X)$$

hint: First prove this result for simple random variables by using the given definition of convexity, then obtain a sequence of simple random variables that converge to the given random variable and take limits using Lebesgue's dominated convergence theorem.

5.9 More Assignment problems in probability theory

[1] Let $\mathbf{X}(n), n \in \mathbb{Z}$ be random walk on the d dimensional lattice, ie, $\mathbf{X}(n) \in \mathbb{Z}^d$ with transition probabilities given by

$$P(\mathbf{X}(n+1)-\mathbf{X}(n) = \mathbf{e}_k|X(n)) = p(k), P(\mathbf{X}(n+1)-\mathbf{X}(n)$$
$$= -\mathbf{e}_k|\mathbf{X}(n)) = q(k), k = 1, 2, ..., d$$

where

$$\mathbf{e}_k = [0, 0, .., 0, 1, 0, ..., 0]^T \in \mathbb{Z}^d$$

with a one in the dth position and zeros at all the other positions and

$$p(k), q(k) \geq 0, \sum_{k=1}^{d}(p(k) + q(k)) = 1$$

Define the probability of the random walk being at the position $\mathbf{k} = \sum_{j=1}^{d} k_j\mathbf{e}_j$ at time n by

$$P(n, \mathbf{k}) = Pr(\mathbf{X}(n) = \mathbf{k})$$

where

$$\mathbf{k} = [k_1, ..., k_d] = k_1\mathbf{e}_1 + .. + k_d\mathbf{e}_d, k_1, ..., k_d \in \mathbb{Z}$$

From elementary intuition, derive the recurrence relation

$$P(n+1, \mathbf{k}) = \sum_{j=1}^{d}(P(n, \mathbf{k}-\mathbf{e}_j)p(j) + P(n, \mathbf{k}+\mathbf{e}_j)q(j))$$

By elementary intuitive arguments, show that if

$$P(n, \mathbf{k}) = \sum_{r_1,...,r_d,s_1,...,s_d} \frac{n!}{r_1!...r_d!s_1!...s_d!} p(1)^{r_1}...p(d)^{r_d}q(1)^{s_1}...q(d)^{s_d}$$

where the sum is over all non-negative integers $r_1, ..., r_d, s_1, ..., s_d$ for which

$$r_j - s_j = k_j, j = 1, 2, ..., d, \sum_{j=1}^{d}(r_j + s_j) = n$$

Now, suppose we view this random walk as a space-time discretized version of a continuous time-stochastic process $\mathbf{Y}(t)$ with values in $\mathbb{R}^d$ such that if $f(t, \mathbf{x})$ is the probability density of $\mathbf{Y}(t)$ with $\mathbf{x} \in \mathbb{R}^d$ and $P(n, \mathbf{k})$ is approximated by $f(n\tau, \mathbf{k}\Delta)\Delta^d$ with τ being the time discretization step size and Δ the spatial discretization step size, then show that the above recursion can be expressed as

$$f(t + \tau, \mathbf{x}) = \sum_{j=1}^{d}(p(j)f(t, \mathbf{x} - \Delta.\mathbf{e}_j) + q(j)f(t - \tau, \mathbf{x} + \Delta\mathbf{e}_j))$$

Show that if

$$\Delta \to 0, \tau \to 0, \Delta^2/2\tau \to D,$$

$$p(j) - q(j) \to 0, (p(j) - q(j))\Delta/\tau \to v_j, p(j) + q(j) \to a(j)$$

then taking this continuum limit, $f(t, \mathbf{x})$ will satisfy the partial differential equation (a diffusion equation)

$$\frac{\partial f(t, \mathbf{x})}{\partial t} = \sum_{j=1}^{d}(-v_j\frac{\partial f(t, \mathbf{x})}{\partial x_j} + D_j\frac{\partial^2 f(t, \mathbf{x})}{\partial x_j^2})$$

where

$$D_j = Da(j), j = 1, 2, ..., d$$

Assuming that at time $t = 0$ the particle executing this diffusion process is located at the origin, ie,

$$f(0, \mathbf{x}) = \delta(\mathbf{x})$$

show that if we define the spatial Fourier transform of f by

$$F(t, \mathbf{K}) = \int f(t, \mathbf{x})exp(i\mathbf{K}.\mathbf{x})d^d\mathbf{x}$$

where

$$\mathbf{K}.\mathbf{x} = \sum_{j=1}^{d} K_j x_j$$

then F satisfies the ode

$$\frac{\partial F(t, \mathbf{K})}{\partial t} = (i\mathbf{v}, \mathbf{K}) + \mathbf{K}^T\mathbf{D}\mathbf{K})F(t, \mathbf{K}), t \geq 0$$

with the initial condition,

$$F(0, \mathbf{K}) = 1$$

where

$$\mathbf{v} = [v_1, ..., v_d]^T, \mathbf{D} = diag[D_1, ..., D_d]$$

so that

$$(\mathbf{v}, \mathbf{K}) = \mathbf{v}^T \mathbf{K} = \sum_{j=1}^{d} v_j K_j,$$

$$\mathbf{K}^T \mathbf{D} \mathbf{K} = \sum_{j=1}^{d} D_j K_j^2$$

Show that the solution is

$$F(t, \mathbf{K}) = exp(it\mathbf{v}^T \mathbf{K} - t\mathbf{K}^T \mathbf{D}\mathbf{K}) = exp(it \sum_{j=1}^{d} K_j v_j - t \sum_{j=1}^{d} D_j K_j^2)$$

which is the characteristic function of a d-dimensional Gaussian random vector having mean $\mathbf{v}t$ and covariance matrix $2t\mathbf{D}$. By Fourier inversion show that

$$f(t, \mathbf{x}) = (4\pi.det(\mathbf{D})dt)^{-1/2}.exp(-(\mathbf{x} - \mathbf{v}t)^T \mathbf{D}^{-1}(\mathbf{x} - \mathbf{v}t)/4t)$$

$$= (4\pi.D_1...D_d t)^{-1/2}.exp(-\sum_{j=1}^{d}(x_j - v_j t)^2/4D_j t)$$

5.10 Multiple choice questions on probability theory

Instructions:Select the most appropriate answer.

[1] Let X, Y be two random variables with joint density $f(x, y)$ and marginal densities $f_X(y), f_Y(y)$ respectively. Then the probability density of $Z = X + Y$ is given by

$$[a] \int f(z + y, y)dy$$

$$[b] \int f_X(z - y) f_Y(y)dy$$

$$[c] \int f(z - y, y)dy$$

$[d] none of the above$

[2] Let X be the outcome when a fair die is thrown. Then, the probability distribution function (CDF) of X is given by

$$[a]\frac{1}{6}\sum_{k=1}^{6}\delta(x-k)$$

$$[b]\frac{1}{6}\sum_{k=1}^{6}\theta(x-k)$$

$$[c]\theta(x)/6$$

$$[d] none of the above.$$

Here, $\theta(x)$ is the unit step function.

[3] Let $X_1, X_2, ..., X_n$ be random variables defined on the same probability space that are non necessarily uncorrelated. Then the variance $Var(S_n)$ of $S_n = X_1 + ... + X_n$ is given by

$$[a]\sum_{k=1}^{n} Var(X_k)$$

$$[b]\sum_{k=1}^{n} Var(X_k) + \sum_{1\leq k<j\leq n} Cov(X_k, X_j)$$

$$[c]\sum_{k=1}^{n} Var(X_k) + 2\sum_{1\leq k<j\leq n} Cov(X_k, X_j)$$

$$[d] none of the above.$$

[4] Let $X_n, n \in \mathbb{Z}$ be a stationary L^1 sequence in a probability space with mean μ and autocovariance $C(\tau)$. Then $|\tau| \to \infty$, and writing $S_n = X_1+...+X_n$, we have that

[a] S_n/n always converges in L^2 to μ

[b] S_n/n converges with probability one to $\mathbb{E}(X_1|\mathcal{T})$ where $\mathcal{T}$ is the tail σ-field.

[c] S_n/n converges in L^2 to $\mathbb{E}X_1$ if $C(\tau)$ is summable.

[d] Both [b] and [c].

[5] Let $(\Omega, \mathcal{F}, P)$ be a probability space and let X be a random variable on this space. Then, the probability distribution of X is given by

[a] $F_X(x) = P(X^{-1}((-\infty, x]))$

[b] $F_X(x) = P(X^{-1}((-\infty, x)))$

[c] $F_X(x) = P(X^{-1}(\{x\})$

[d] None of the above

[6] Let $X(t)$ be an L^2 stochastic process that is passed through a time varying filter with impulse response $h(t,s)$ so that the filter output is given by $Y(t) = \int h(t,\tau)X(\tau)d\tau$. Then if $R_{XX}(t,\tau) = \mathbb{E}(X(t)X(\tau))$, $R_{YY}(t,\tau) = \mathbb{E}(Y(t)Y(\tau))$ and $R_{YX}(t,\tau) = \mathbb{E}(Y(t)X(\tau))$, we have

[a] $R_{YY}(t,\tau) = \int h(t,s)R_{XX}(s,\tau)ds$
[b] $R_{YY}(t,\tau) = \int h(t,s)R_{YX}(s,\tau)ds$
[c] $R_{YY}(t,\tau) = \int h(t,s_1)h(\tau,s_2)R_{XX}(s_1,s_2)ds_1ds_2$
[d] Both [b] and [c]

[7] Let $(\Omega_k,\mathcal{F}_k), k = 1,2$ be two measurable spaces and let $T : (\Omega_1,\mathcal{F}_1) \to (\Omega_2,\mathcal{F}_2)$ be a measurable transformation. P_1 be a probability measure on the first space and $P_2 = P_1 o T^{-1}$. Let X be an integrable random variable on the second probability space. Then,

[a] $\int X(\omega)dP_2(\omega) = \int X(T\omega)dP_1(\omega)$
[b] $\int X(\omega)dP_1(\omega) = \int X(T\omega)dP_2(\omega)$
[c] $\int X(\omega)dP_1(\omega) = \int X(\omega)dP_2(\omega)$
[d] None of the above.

[8] Let $\{X_n\}$ be a sequence of random variables defined on a probability space. Then consider the following statements: [1] The probability distribution of X_n converges to the probability distribution F of a r.v X at all continuity points of F, [2] $P(|X_n - X| > \epsilon) \to 0$ for all $\epsilon > 0$, [3] $\mathbb{E}(X_n - X)^2 \to 0$, [4] $P(X_n \to X) = 1$. Then

[a] [3] $\implies$ [2]
[b] [4] $\implies$ [2] $\implies$ [1]
[c] [4] $\implies$ [3]
[d] both [a] and [b]

[9] Let $X(t)$ be a random process passed through a system having the input output relation described by the differential equation

$$d^2Y(t)/dt^2 = adY(t)/dt + bY(t) + X(t)$$

Then the cross-correlation $R_{YX}(t,s) = \mathbb{E}(Y(t)X(s))$ satisfies

[a] $\frac{\partial^2 R_{YX}(t,s)}{\partial t^2} = a\frac{\partial R_{YX}(t,s)}{\partial t} + bR_{YX}(t,s) + R_{XX}(t,s)$
[b] $\frac{\partial^2 R_{YX}(t,t)}{\partial t^2} = a\frac{\partial R_{YX}(t,t)}{\partial t} + bR_{YX}(t,t) + R_{XX}(t,t$
[c] $\frac{\partial^2 R_{YX}(t,s)}{\partial t^2} = a\frac{\partial R_{YY}(t,s)}{\partial t} + bR_{YY}(t,s) + R_{XX}(t,s)$
[d] None of the above.

[10] Let X be a random variable with a strictly increasing probability distribution function $F(x)$. Then, if U is a uniformly distributed random variable with values in $[0,1]$, the random variable $F^{-1}(U)$ has the probability distribution

[a] uniform

[b] $F(x)$
[c] $F^{-1}(x)$
[d] none of the above.

5.11 Design of a quantum unitary gate using superstring theory with noise analysis based on the Hudson-Parthasarathy quantum stochastic calculus

The superstring action is given by

$$S[X,\psi] = \int ((1/2)\partial_\alpha X^\mu \partial^\alpha X_\mu - i\psi^{\mu T}\sigma^2\rho^\alpha\partial_\alpha\psi_\mu)d^2\sigma + \int B_{\mu\nu}(X)dX^\mu \wedge dX^\nu$$

Note that

$$dX^\mu \wedge dX^\nu = (X^\mu_{,1}X^\nu_{,2} - X^\mu_{,2}X^\nu_{,1})d^2\sigma$$
$$= \epsilon^{\alpha\beta}\partial_\alpha X^\mu.\partial_\beta X^\nu$$

where

$$\epsilon^{12} = 1, \epsilon^{21} = -1, \epsilon^{11} = \epsilon^{22} = 0$$

The supersymmetry transformations under which this action is invariant are

$$\delta X^\mu = c_1 k^T\sigma^2\psi^\mu,$$
$$\delta\psi^\mu = c_2\rho^\alpha k\partial_\alpha X^\mu$$

Here,

$$\rho^0 = \sigma^1, \rho^1 = \sigma^3$$

where k is an infinitesimal Fermionic parameter. This supersymmetric action can also be derived using basic superfield theory by defining our superfield on the space of two dimensional Bosonic and two dimensional Fermionic space as

$$\Phi^\mu(\sigma,\theta) = X^\mu(\sigma) + \theta^T\epsilon\psi^\mu(\sigma) + \theta^T\epsilon\theta.Y$$

where Y is a scalar Bosonic field and $\epsilon = i\sigma^2$. The infinitesimal supersymmetry transformations are defined by the super-vector field

$$L = k^T(\epsilon\rho^\alpha\theta.\partial_\alpha + \partial/\partial\theta)$$

with k being an infinitesimal Fermionic parameter. It is clear that under such an infinitesimal transformation of the superfield Φ, we have

$$\delta X^\mu = k^T\epsilon\psi^\mu,$$
$$\theta^T\epsilon\delta\psi^\mu = k^T\epsilon\rho^\alpha\theta.\partial_\alpha X^\mu$$

or equivalently,

$$\delta\psi^\mu = \rho^\alpha k\partial_\alpha X^\mu$$

5.12 Study projects in probability theory:Construction of Brownian motion, law of the iterarted logarithm

[1] Construction of Brownian motion on $[0,1]$ using the Haar basis.

Step 1: For $n \geq 1$ and $k = 0, 1, ..., 2^{n-1} - 1$, define $H_{n,k}(t)$ to be $2^{(n-1)/2}$ for $t \in [2k/2^n, (2k+1)/2^n)$ and $-2^{(n-1)/2}$ for $t \in [(2k+1)/2^n, (2k+2)/2^n)$. For all other $t \in [0,1]$ set $H_{n,k}(t) = 0$. Define $H_0(t) = 1$. Show that if $f \in L^2[0,1]$ and $< f, H_{n,k} >, < f, H_0 >= 0$ for all n, k, then $f = 0$. This proved by noting that $f \perp H_{n,k}$ implies

$$\int_{2k/2^n}^{(2k+1)/2^n} f(t)dt = \int_{(2k+1)/2^n}^{(2k+2)/2^n} f(t)dt, k \in I(n)$$

where

$$I(n) = \{0, 1, ..., 2^{n-1} - 1\}$$

By considering the limit of the above equations as $n \to \infty$, deduce that $f \perp H_{n,k}$ for all n, k implies that f is a.s a constant on $[0,1]$ and by further making use of $f \perp H_0$, deduce that this constant is zero.

Step 2: Prove the orthogonality relations

$$< H_{nk}, H_{ml} >= \delta_{nm}\delta_{kl}$$

Do this by first taking $m = n$ and noting then that if $k \neq l$, then H_{nk}, H_{ml} have disjoint supports, ie non-overlapping supports and hence the two are orthogonal. Next observe that if $n > m$ then by expressing $j/2^m$ as $j.2^{n-m}/2^n$, it follows that if H_{nk} and H_{ml} have overlapping supports, then the support of H_{nk} is contained entirely in the first half or entirely in the second half of that of H_{ml} and hence by using the fact that the integral of H_{nk} is zero, deduce orthogonality of these two. Finally observe the trivial result that H_{nk} is orthogonal to H_{00} since its integral over $[0,1]$ is zero. Conclude then that $H_{00}, \{H_{nk} : n \geq 1, k \in I(n)\}$ is an onb for $L^2[0,1]$.

Step 3: Define the Schauder functions

$$S_{nk}(t) = \int_0^t H_{nk}(s)ds$$

and using the identity

$$f(t) = \sum_{n,k} < f, H_{nk} > H_{nk}(t), f \in L^2[0,1]$$

deduce by taking $f(s) = \chi_{[0,t]}(s), t, s \in [0,1]$ that

$$\sum_{n,k} S_{nk}(t)S_{nk}(s) = min(t, s)$$

Observe that the graph of $S_{nk}(t)$ is a symmetric tent of height $2^{-(n+1)/2}$ over the interval $[(2k-1)/2^n, (2k+2)/2^n]$, the symmetry being about the mid point of this interval $(2k+1)/2^n$. Observe also that for a fixed n, the $S'_{nk}s$ have non-overlapping supports as k varies over $I(n)$.

Step 4: Now let $\xi(n,k), n \geq 1, k \in I(n)$ be iid $N(0,1)$ r.v's. Define

$$b(n) = max(|\xi(n,k)| : k \in I(n)\}$$

and observe that

$$P(b(n) > n) \leq 2^{n-1}.P(|\xi(1,1)| > n) = 2^n.(1-\Phi(n)) \leq 2^n.exp(-n^2/2)/n\sqrt{2\pi}$$

and hence

$$\sum_n P(b(n) > n) < \infty$$

so that by the Borel-Cantelli lemma,

$$P(b(n) > n, i.o) = 0$$

Show that this is the same as saying that for a.e.ω, there exists a finite positive integer $N(\omega)$ such that

$$b(n,\omega) \leq n, \forall n > N(\omega)$$

Conclude that

$$\sum_{k\in I(n)} |\xi(n,k)(\omega)S_{nk}(t)| \leq b(n,\omega) \sum_{k\in I(n)} S_{nk}(t)$$

$$\leq b(n,\omega)2^{-(n+1)/2} \leq n.2^{-(n+1)/2}, \forall n > N(\omega)$$

Conclude that if we define the processes

$$B_N(t) = \sum_{n=1}^{N} \sum_{n\in I(n)} \xi(n,k)S_{nk}(t), N \geq 1$$

then the $B'_N s$ are continuous processes that converge uniformly to a limiting process $B(t)$ over $[0,1]$ almost surely and hence the limiting process $B(t)$ has almost surely continuous sample paths. Complete the proof by showing that

$$\mathbb{E}B_N(t) = 0, Cov(B_N(t), B_N(s)) = \sum_{k\in I(n), 1\leq n\leq N} S_{nk}(t)S_{nk}(s)$$

and further that for each t $B_N(t)$ is a Cauchy sequence in $L^2(\Omega, \mathcal{F}, P)$ and therefore converges in L^2 to $B(t)$. Then using continuity properties of the inner product in $L^2(\Omega, \mathcal{F}, P)$, deduce that

$$\mathbb{E}(B(t)) = lim_N\mathbb{E}(B_N(t)), \mathbb{E}(B(t)B(s)) = lim_N\mathbb{E}(B_N(t)B_N(s))$$

$$= \sum_{n\geq 1, k\in I(n)} S_{nk}(t)S_{nk}(s) = min(t,s)$$

ie, $B(.)$ is a Gaussian process with a.s. continuous sample paths having zero mean and covariance $min(t, s)$, or in other words, $B(.)$ is a Brownian motion process over $[0, 1]$.

Remark: To prove that $B(.)$ is a Gaussian process, it suffices to show that if $\{t_1, ..., t_k\}$ is a finite set of points in $[0, 1]$, then $(B(t_1), ..., B(t_k))$ is a Gaussian random vector. But this is immediately a consequence of the fact that $(B_N(t_1), ..., B_N(t_k))$ is a Gaussian random vector which converges in distribution since it converges in probability since it converges a.s since the process $B_N(.)$ converges uniformly a.s.

[2] The law of the iterated logarithm: This result states that if $B(.)$ is standard BM, then

$$limsup_{t\to\infty}\frac{B(t)}{\sqrt{2t.loglog(t)}} = 1 a.s$$

Equivalently, since $tB(1/t)$ is also a BM, we can state the law of the iterated logarithm as

$$limsup_{t\to 0}\frac{B(t)}{2t.loglog(1/t)} = 1 a.s$$

Equivalently since $-B(t)$ is also a BM, this law can also be stated as

$$liminf_{t\to\infty}\frac{B(t)}{2t.loglog(t)} = -1$$

Intuitively what these result state is that for very large times t, $B(t)$ almost surely oscillates between the two bounding curves $x = \pm\sqrt{2t.loglog(t)}$.

To prove this result, define

$$h(t) = \sqrt{2t.lnln(1/t)}, 0 < t < 1$$

so that for $0 < \theta < 1$ and $n = 0, 1, ...,$ we have

$$h(\theta^n) = \sqrt{2\theta^n.ln(n.ln(1/\theta)}$$

Then we use Doob's Martingale's inequality in the form

$$P(max_{0<s<t}(B(s)-\lambda s/2) > \beta) = P(max_{0<s<t}exp(\lambda.B(s)-\lambda^2 s/2) \leq exp(-\lambda\beta))$$

$$\leq exp(-\lambda\beta), \lambda > 0$$

In this inequality, we choose

$$\beta = h(\theta^n)/2, \lambda = (1+\delta)\theta^{-n}h(\theta^n)$$

to get

$$P(max_{\theta^{n+1}<s<\theta^n}(B(s)-(1+\delta)\theta^{-n}h(\theta^n)s/2) > h(\theta^n)) \leq exp(-(1+\delta)\theta^{-n}h(\theta^n)^2/2)$$

from which we deduce that

$$P(max_{\theta^{n+1}<s<\theta^n} B(s) > h(\theta^n))(1+\delta/2))$$

$$\leq exp(-(1+\delta)ln(nln(1/\theta)) = O(n^{-(1+\delta)})$$

and the rhs is summable. So by the Borel-Cantelli lemma and using the fact that for $t \in (\theta^{n+1}, \theta^n)$, we have

$$h(t) = \sqrt{2t.lnln(1/t)} \geq \sqrt{2.\theta^{n+1}.lnln(1/\theta^n)}$$

$$= \sqrt{\theta}h(\theta^n)$$

and we get

$$P(max_{\theta^{n+1}<s<\theta^n}(B(s)/h(s)) > (1+\delta/2)\sqrt{\theta}, i.o) = 0$$

Thus,

$$P(limsup_{s\to 0}(B(s)/h(s)) > (1+\delta/2)\sqrt{\theta}) = 0$$

Letting θ increase to one through a set of rationals gives us

$$P(limsup_{s\to 0}(B(s)/h(s)) > (1+\delta/2)) = 0 \forall \delta > 0$$

and letting now $\delta \downarrow 0$ gives us the result that

$$P(limsup_{s\to 0}(B(s)/h(s) \leq 1) = 1$$

This completes the proof of the first half of the law of the iterated logarithm. For the second half, we consider the events

$$E_n = \{(B(\theta^n) - B(\theta^{n+1})/\sqrt{\theta^n - \theta^{n+1}} > x_n)$$

$$= 1 - \Phi(x_n))$$

where Φ is the standard normal distribution. We choose

$$x_n = \sqrt{1-\theta}h(\theta^n)/\sqrt{\theta^n - \theta^{n+1}} = h(\theta^n)/(\theta^{n/2})$$

We have for $x > 0$, the inequality

$$1 - \Phi(x) = (2\pi)^{-1/2} \int_x^\infty exp(-y^2/2)dy$$

$$=\geq (2\pi)^{-1/2} \int_x^\infty (y/x)exp(-y^2/2)dy$$

$$= (2\pi)^{-1/2} xexp(-x^2/2), x > 0$$

In particular,

$$(1 - \Phi(x_n)) \geq (2\pi)^{-1/2}(h(\theta^n)/\theta^{n/2})exp(-h(\theta^n)^2/2\theta^n)$$

$$= K\sqrt{2ln(n.ln(1/\theta))}.exp(-ln(n.ln(1/\theta)))$$

whose sum over $n \geq 1$ is clearly divergent. Hence, by the second Borel-Cantelli lemma, the events

$$E_n = \{(B(\theta^n) - B(\theta^{n+1}))/h(\theta^n) > \sqrt{1-\theta}\}, n \geq 1$$

occur infinitely often with probability one. Further, from the previous half with the Brownian motion B replaced by $-B$, we have the result that a.s. there exists an integer $N = N(\omega)$ such that

$$\{-B(\theta^{n+1})/h(\theta^{n+1}) \leq 1 + \delta, \forall n > N$$

This is the same as saying that

$$-B(\theta^{n+1})/h(\theta^n) \leq \sqrt{\theta}(1+\delta)\forall n > N$$

and combining this with the previous result gives us the result that the events

$$F_n = \{B(\theta^n)/h(\theta^n) > \sqrt{1-\theta} - (1+\delta)\sqrt{\theta}\}$$

occur infinitely often. This true for every $\delta > 0$ and for every $\theta \in (0,1)$. Letting first $\delta \downarrow 0$ gives us the result that

$$P(limsup_{t\to 0}B(t)/h(t) > \sqrt{1-\theta} - \sqrt{\theta}) = 1$$

Then letting $\theta \uparrow 1$ gives us the result that

$$P(limsup_{t\to 0}B(t)/h(t) \geq 1) = 1$$

and this completes the proof of the law of the iterated logarithm for Brownian motion:

$$P(limsup_{s\to 0}B(s)/h(s) = 1) = 1$$

5.13 Quantum Boltzmann equation for a system of particles interacting with a quantum electromagnetic field

Let $\rho(t) = \rho_{123..N}(t)$ denote the state of the system of N identical particles. This state is an operator on $\mathcal{H}^{\otimes N}$ and it satisfies Schrodinger's equation

$$i\rho'(t) = [H(t), \rho(t)]$$

where

$$H(t) = \sum_{j=1}^{N}((p_j + eA(t, r_j))^2/2m - e\Phi(t, r_j)) + H_F(t)$$

where $H_F(t)$ is the electromagnetic field Hamiltonian in Boson Fock space. It is given by

$$H_F(t) = (\epsilon/2)\int |E(t,r)|^2 d^3r + (1/2\mu)\int |B(t,r)|^2 d^3r$$

where

$$E(t,r) = -\nabla\Phi(t,r) - \partial_t A(t,r), B(t,r) = curlA(t,r)/\mu$$

The Maxwell equations for E, B are written down taking into account the quantum current density and charge density associated with the charges of the N particles and their joint density operator $\rho(t)$. If $\rho_1(t,r,r')$ denotes the position space representation of the marginal density for one particle, then we know by analogy with the expression for the quantum current and charge density in a pure state $\psi(t,r)$,

$$\mathbf{J}(t,r) = (i/2m)(\psi(t,r)^*\nabla\psi(t,r) - \psi(t,r)\nabla\psi(t,r)^*),$$

$$\sigma(t,r) = \psi(t,r)^*\psi(t,r)$$

that the same quantities in the mixed state ρ_1 are given by

$$\mathbf{J}(t,r) = (i/2m)[\nabla_1\rho(t,r,r') - \nabla_2\rho(t,r,r'))]|_{r'=r}$$

$$\sigma(t,r) = \rho(t,r,r)$$

Note that

$$\rho_1(t) = Tr_{23...N}\rho_{123...N}(t)$$

These expressions for the current and charge densities are to be substituted into the Maxwell equations

$$curlE(t,r) = -\partial_t B(t,r), curlB(t,r) = \mu J(t,r) + \mu\epsilon\partial_t E(t,r),$$

$$divB(t,r) = 0, divE(t,r) = \sigma(t,r)/\epsilon$$

This model can be used to describe a quantum plasma within a quantum cavity resonator having a quantum electromagnetic field within it. The solutions for the quantum electromagnetic field will be given by a sum of two terms:The first term is the free field solutions as conventionally described in quantum electrodynamics in terms of the photon creation and annihilation operators. This part is the solution to the homogeneous (ie, source free) part of the Maxwell equations. The second term is the particular solution of the Maxwell equations that is linear in the current and charge densities. This expression for the electromagnetic field operators is to be substituted into the quantum Boltzmann equation for $\rho_1(t,r,r')$ in order to get an appropriate description of the plasma.

5.14 Device physics in a semiconductor using the classical Boltzmann kinetic transport equation

Study project. Study the effect of small electric and magnetic field perturbations on the solution to the kinetic transport equation using first order perturbation theory and hence evaluate approximately the current density produced by such field fluctuations. Use this to describe the photocurrent in a device when radiation falls upon it.

5.15 A.Describing the value of a point charge and its location in space in terms of the electrostatic potential generated by it

B.Rewriting Dirac's relativistic wave equation for the electron interacting with a quantum electromagnetic field and the classical nuclear field without explicitly bringing in the electronic charge

Let the point charge be Q and let its location be r_0. The potential produced by it is

$$V(r) = Q/4\pi\epsilon|r - r_0|$$

Thus,

$$\nabla^2 V(r) = -Q\delta(r - r_0)/\epsilon$$

We can thus recover Q and r_0 from $V(.)$ using the formula

$$\int f(r)\nabla^2 V(r)d^3r = (-Q/\epsilon)f(r_0)$$

for any measurable function f having compact support. In particular, let $g(r)$ be another function. Then,

$$\frac{\int f(r)\nabla^2 V(r)d^3r}{\int g(r)\nabla^2 V(r)d^3r} = f(r_0)/g(r_0)$$

Therefore, if f, g are functions that are zero outside a compact subset K of $\mathbb{R}^3$ and are such that $f(x, y, z)/g(x, y, z) = x$, then

$$x_0 = \frac{\int f(r)\nabla^2 V(r)d^3r}{\int g(r)\nabla^2 V(r)d^3r}$$

Likewise y_0, z_0 can be recovered from $V(.)$. In this way, we can recover $r_0 = (x_0, y_{0,0})$ from $V(.)$ Then, Q is also determined using

$$Q = (-\epsilon/f(r_0))\int f(r)\nabla^2 V(r)d^3r$$

Now consider the problem of determining the point charges and their locations given their number from the electrostatic potential generated by them. Let these charges be $Q_1, ..., Q_n$ and let their locations be $r_1, ..., r_n$. If $V(r)$ is the electrostatic potential generated by them, then Poisson's equation gives

$$\epsilon \nabla^2 V(r) = \sum_{k=1}^{n} Q_k \delta(r - r_k)$$

and hence, if $f(r)$ is a bounded measurable function, we have that

$$\sum_{k=1}^{n} Q_k f(r_k) = \epsilon \int f(r) \nabla^2 V(r) d^3 r$$

Now choose the functions $f_1(r) = x^m, f_2(r) = y^m, f_3(r) = z^m, m = 0, 1, 2, ...$ and derive from the above, the following system of equations

$$\sum_{k=1}^{n} Q_k x_k^m = \int x_k^m \nabla^2 V(r) d^3 r,$$

$$\sum_{k=1}^{n} Q_k y_k^m = \int y_k^m \nabla^2 V(r) d^3 r,$$

$$\sum_{k=1}^{n} Q_k z_k^m = \int z_k^m \nabla^2 V(r) d^3 r$$

for $m = 0, 1, 2, ...N - 1$. Thus we get a system of $3N$ equations which can be solved for $Q_k, x_k, y_k, z_k, k = 1, 2, ..., n$ or a least squares solution can be obtained provided that $N \geq 4n$. In this way a finite discrete charge distribution in space can be completely determined from the potential field. We could also do this using measurements of the electric field only using Gauss' law:

$$\epsilon div E(r) = \sum_{k=1}^{n} Q_k \delta(r - r_k)$$

Thus,

$$\epsilon \int f(r) div E(r) d^3 r = \sum_{k=1}^{n} Q_k f(r_k)$$

Another way to express the point charge distribution as a functional of the potential is to assume that the distance between the locations of any two charges in the set is greater than 2δ. Let $Q_1, ..., Q_n$ denote the point charges with locations $r_1, ..., r_n$ so that the charge density is

$$\rho(r) = \sum_{k=1}^{n} Q_k \delta(r - r_k)$$

with

$$|r_k - r_j| > \delta \forall k \neq j$$

Then let $B(\delta)$ denote the open ball in $\mathbb{R}^3$ with the origin as centre and radius δ:

$$B(\delta) = \{r \in \mathbb{R}^3 : |r| < \delta\}$$

Then given an arbitary point $r \in \mathbb{R}^3$, we have that $B(r, \delta) = r + B(\delta)$ can contain at most only one of the r_k's. It follows that for any r,

$$-\epsilon \int_{B(r,\delta)} \nabla^2 V(r') d^3 r'$$

equals either zero or Q_k for some $k = 1, 2, ..., n$. It equals Q_k iff $|r - r_k| < \delta$. In other words, by moving the center ball $B(\delta)$ to different points, we get a result either equal to zero or Q_k and from the location of the centre of the ball, we can determine r_k upto an accuracy of δ. This result can also be stated as

$$lim_{r \to r_k, \delta \to 0} \int_{B(r,\delta)} (-\epsilon \nabla^2 V(r')) d^3 r'$$

$$= Q_k$$

Now consider an electron of charge $-e$ interacting with the atomic nucleus of charge Ze. Let $A_q(t, r)$ denote the free quantum electromagnetic field in space-time and let $\psi(t, r)$ denote the second quantized Dirac wave function of the electron. It satisfies the equation

$$[(\gamma^\mu (i\partial_\mu + eA_{q\mu}(t, r) + eA_{N\mu}(r)) - m]\psi(t, r) = 0$$

where

$$A_{N0}(r) = -Ze^2/|r|, A_{Nj}(r) = 0, j = 1, 2, 3$$

is the classical nuclear potential. According to our theory, A_q, the free quantum electromagnetic field does not have any singularity and hence if δ is sufficiently small, we have

$$\epsilon \int_{B(\delta)} \nabla^2 (A_{q\mu}(t, r) + A_{N\mu}(r)) d^3 r = -Ze\delta_{\mu,0}$$

This equation then determines the electron charge $-e$ from the **total** electromagnetic potential. One way to write down Dirac's equation without introducing explicitly the electronic charge $-e$ is then to write it as

$$[\gamma^\mu (i\partial_\mu - Z^{-1}[\int_{B(\delta)} \nabla^2 A_0(t, r') d^3 r')] A_\mu(t, r)]) - m]\psi(t, r) = 0 - - - (1)$$

where

$$A_\mu = A_{qmu} + A_{N\mu}$$

is the total electromagnetic four potential comprising of the free field part described in term of photon creation and annihilation operators and the classical nuclear part.

Remark: It should be noted that in our formalism, the electron exists only because it is a part of the atom having an atomic nucleus. The existence of the electron without a corresponding nucleus is not meaningful.

Now the total electromagnetic field A_μ in the region $|r| > 0$, ie, in $\mathbb{R}^3 - \{0\}$ satisfies the wave equation

$$\nabla^2 A_\mu - \frac{1}{c^2}\partial_t^2 A_\mu = 0, \partial^\mu A_\mu = 0 - - - (2)$$

The question is , "Is it possible to derive all the consequences of conventional quantum field theory from equns (1) and (2) only ? To answer this question, let us assume first that the total electromagnetic field A_μ is given and we have to solve (1). Perturbatively solving it gives us to a first order approximation in the interaction term

$$\psi = \psi_0 + \psi_1,$$

$$[i\gamma^\mu\partial_\mu - m]\psi_0 = 0,$$

$$[i\gamma^\mu\partial_\mu - m]\psi_1 =$$

$$Z^{-1}[\int_{B(\delta)} \nabla^2 A_0(t,r')d^3r')]A_\mu(t,r)\gamma^\mu\psi_0$$

ψ_0 therefore represents the free Dirac field expressible as a superposition of electron and positron creation and annihilation operators. The first order perturbation ψ_1 to the free Dirac field is then

$$\psi_1 = Z^{-1}\int S(x-x')[\int_{B(\delta)} \nabla^2 A_0(t,r')d^3r')]A_\mu(x')\gamma^\mu\psi_0(x')d^4x' - - - (3)$$

We now use (3) to compute radiative corrections to the electron propagator in terms of the photon propagator:

$$< T(\psi(x)\psi(x')^*) >\approx< T(\psi_0(x)\psi_0(x')^*) > + < T(\psi_0(x)\psi_1(x')^*) >$$

$$+ < T(\psi_1(x)\psi_0(x')^*) >$$

where now

$$< T(\psi_0(x)\psi_0(x')^*) >= S_0(x-x')$$

is the free Dirac field electron propagator known to be given by

$$S_0(x-x') = K.\int(\gamma^\mu p_\mu - m)^{-1}exp(ip.x)d^4p$$

Now come to the time varying case for calculating charges and their velocities from the singularities in the electromagnetic field. A point charge Q

moving along the trajectory $R(t), t \geq 0$ with non-relativistic velocity generates an electromagnetic field given approximately by

$$E(t,r) = \frac{Q(r-R(t))}{4\pi\epsilon|r-R(t)|^3},$$

$$B(t,r) = \frac{\mu QV(t) \times (r-R(t))}{|r-R(t)|^3}$$

Equivalently in terms of the Maxwell equations,

$$divE(t,r) = Q\delta(r-R(t))/\epsilon,$$

$$curlB(t,r) = \mu QV(t)\delta(r-R(t)) + \epsilon\partial_t E(t,r)$$

and hence we deduce that for a smooth function $f(r)$ of space coordinates,

$$\int f(r)divE(t,r)d^3r = (Q/\epsilon)f(R(t)),$$

$$\int f(r)curlB(t,r)d^3r - \epsilon\int f(r)\partial_t E(t,r)d^3r = \mu QV(t)f(R(t))$$

and by selecting f appropriately, it is clear from these two equations how to determine the charge and its trajectory including velocity at each time from the total electromagnetic field in space.

We apply this idea to quantize the electromagnetic field and Dirac field when the nucelus having charge $Q = Ze$ that binds the electron moves along a trajectory $R(t)$. The magnetic vector potential generated by such a nucleus is

$$A_N(t,r) = \frac{\mu QV(t)}{4\pi|r-R(t)|},$$

and the electric scalar potential is given by

$$A_{N0}(t,r) = \frac{Q}{4\pi\epsilon|r-R(t)|}$$

in the non-relativistic approximation. Equivalently, in the non-relativistic approximation, we have

$$\nabla^2 A_N(t,r) = -\mu QV(t)\delta(r-R(t)),$$

$$\nabla^2 A_{N0}(t,r) = -Q\delta(r-R(t))/\epsilon$$

so that for any test function $f(r)$, we have

$$\int f(r)\nabla^2 A_N(t,r)d^3r = -\mu QV(t)f(R(t)),$$

$$\int f(r)\nabla^2 A_{N0}(t,r)d^3r = -Qf(R(t))/\epsilon$$

By taking the ratio of these two equations, we obtain the charge velocity vector

$$V(t) = (\mu\epsilon)^{-1} \frac{\int f(r)\nabla^2 A_N(t,r)d^3r}{\int f(r)\nabla^2 A_{N0}(t,r)d^3r}$$

The charge Q can be calculated in terms of the field A_{N0} by integrating the Laplacian applied to it over a small neighbourhood of its position $R(t)$. However, to do so, we require first to estimate $R(t)$ from the field. That can be done by taking $f(r) = |r|$ giving thereby

$$\int |r|.\nabla^2 A_{N0}(t,r)d^3r = -Q|R(t)|/\epsilon,$$

and then taking $f(r) = |r|^2$, we get

$$\int |r|^2\nabla^2 A_{N0}(t,r)d^3r = -Q|R(t)|^2/\epsilon$$

Eliminating $|R(t)|$ between these two equations gives us the charge as

$$Q = -\epsilon \frac{(\int |r|.\nabla^2 A_{N0}(t,r)d^3r)^2}{\int |r|^2\nabla^2 A_{N0}(t,r)d^3r}$$

$R(t)$ may now be calculated using

$$\int r\nabla^2 A_{N0}(t,r)d^3r = -QR(t)/\epsilon$$

and $V(t)$ using

$$\int \nabla^2 A_N(t,r)d^3r = -\mu QV(t)$$

If we assume that the quantum electromagnetic field $A_{q\mu}$ fluctuates rapidly in space, then its spatial average over any small open ball of finite radius will be negligible and hence we can to a good degree of approximation write

$$\int f(r)(\frac{\int_{B(r,\delta)} \nabla^2 A_0(t,r')d^3r'}{V(B(\delta))})d^3r = -Qf(R(t))/\epsilon,$$

$$\int f(r)(\frac{\int_{B(r,\delta)} \nabla^2 A(t,r')d^3r'}{V(B(\delta))})d^3r = -\mu QV(t)f(R(t))$$

where

$$A = A_N + A_q, A_0 = A_{N0} + A_{q0}$$

or equivalently,

$$A_\mu = A_{N\mu} + A_{q\mu}$$

These are respectively the total magnetic vector potential due to the nucleus and the quantum field and the total electrostatic field due to the same. Note that works because the nuclear potential has a singularity at the origin and at other spatial points, it varies slowly in space, while the quantum field is smooth thereby ensuring that

$$\frac{\int_{B(r,\delta)} A_{q\mu}(t,r')d^3r'}{V(B(\delta))} \approx 0$$

and since

$$\frac{\int_{B(r,\delta)} A_{N\mu}(t,r')d^3r'}{V(B(\delta))} \approx A_{N\mu}(t,r)$$

for small positive δ. Therefore,

$$\frac{\int_{B(r,\delta)} A_{\mu}(t,r')d^3r'}{V(B(\delta))} \approx A_{N\mu}(t,r)$$

Dirac's equation for the electron wave function is now expressible entirely in terms of the total electromagnetic field without even bringing in the electronic charge parameter. Formally, this equation is therefore expressible as

Relativistic considerations:Suppose that the nucleus is moving with relativistic velocities. Then, we replace the Laplacian operator by the wave operator in the above equations thereby obtaining

$$\Box A_N(t,r) = -\mu QV(t)\delta(r-R(t)),$$

$$\Box A_{N0}(t,r) = -Q\delta(r-R(t))/\epsilon$$

where

$$\Box = \nabla^2 - \mu\epsilon\partial_t^2$$

It is then clear how all the parameters of the moving nucleus, namely it charge, position trajectory and velocity can be computed as functions of weighted integrals of the total electromagnetic field. Specifically, we find that

$$\int f(r)(\frac{\int_{B(r,\delta)} \Box A(t,r')d^3r'}{V(B(\delta))}d^3r = -\mu QV(t)f(R(t))$$

$$\int f(r)(\frac{\int_{B(r,\delta)} \Box A_0(t,r')d^3r'}{V(B(\delta))}d^3r = -Qf(R(t))/\epsilon$$

The Dirac equation is now of the form

$$[\gamma^\mu(i\partial_\mu + F(A_{nu}(t,r), r \in \mathbb{R}^3)A_\mu(t,r)) - m]\psi(t,r) = 0$$

where $e = F(A_\nu(t,r), r \in \mathbb{R}^3)$ is the electronic charge value determined as above as spatial functional of the electromagnetic field. The electromagnetic field on the other hand satisfies Maxwell's equations in the form

$$\Box A_\mu(t,r) = 0, r \neq R(t)$$

The whole point of this exercise is that by measuring data about the Dirac wave function, or equivalently the Dirac four current density, we can in principle calculate the electromagnetic field A_μ and hence from the singularity theory mentioned above, calculate the nuclear charge as well as its trajectory.

5.16 Calculating the masses of N gravitating particles and their positions and their trajectories from measurement of the gravitational potential distribution in space-time using the Newtonian theory

Let $m_1, ..., m_N$ denote the masses of N point particles moving under their mutual gravitation along trajectories $r_1(t), ..., r_N(t)$. The Newtonian equations of motion are

$$r_j''(t) = \sum_{k=1,k\neq j}^{N} Gm_k(r_j - r_k)/|r_k - r_k|^3, j = 1, 2, ..., N$$

The gravitational potential generated by these masses is then

$$\Phi(r) = \sum_{j=1}^{N} Gm_j/|r - r_j|$$

and this potential satisfies Poisson's equation

$$\nabla^2\Phi(t, r) = 4\pi G \sum_{j=1}^{N} m_j\delta(r - r_j(t))$$

Thus, we get for a test function $f(r)$,

$$\int f(r)\nabla^2\Phi(t, r)d^3r = 4\pi G \sum_{j=1}^{N} m_j f(r_j(t))$$

By choosing test functions $f_1, ..., f_N$ appropriately, we get the following linear system of equations for the masses given their positions:

$$\int_{j=1}^{N} f_k(r_j(t))m_j = \int f_k(r)\nabla^2\Phi(t, r)d^3r, k = 1, 2, ..., N$$

Define the $N \times N$ matrix valued function of time

$$A(t) = ((f_k(r_j(t))))_{1\leq k,j\leq N}$$

and let

$$B(t) = A(t)^{-1} = ((b_{ij}(t)))$$

Then,

$$m_k = \sum_{j=1}^{N} b_{kj}(t) \int f_j(r)\nabla^2\Phi(t, r)d^3r$$

This formula will work even if the masses are functions of time. Now the $b_{jk}(t)'s$ are functions of the $r_j(t)'s$. So the $r_j(t)'s$ also have to be estimated from the potential distribution. Choose N vectors $\xi_1, ..., \xi_N$ in $\mathbb{R}^3$. Then, we have

$$\int < \xi_k, r > \nabla^2\Phi(t,r)d^3r = 4\pi G\sum_{j=1}^N m_j < \xi_k, r_j >, k = 1,2,...,N$$

This is a system of N linear equations for the N masses and defining the matrix

$$((c_{kj}(r_j)))_{1\le k,j\le N} = C(r_1,...,r_N) = 4\pi G((< \xi_k, r_j >))_{1\le k,j\le N}$$

gives us

$$m_k = \sum_{j=1}^N e_{kj}(r_1,...,r_N)\int < \xi_j, r > \nabla^2\Phi(t,r)d^3r, k = 1,2,...,N$$

where

$$((e_{kj})) = C^{-1}$$

Thus we obtain the following N equations for $r_1, ..., r_N$:

$$\sum_{j=1}^N b_{kj}(t)\int f_j(r)\nabla^2\Phi(t,r)d^3r$$

$$= \sum_{j=1}^N e_{kj}(r_1,...,r_N)\int < \xi_j, r > \nabla^2\Phi(t,r)d^3r, k = 1,2,....,N$$

and by varying the vectors ξ_k in these equations, we can derive at least $3N$ equations for the N vectors $r_1, ..., r_N$ which can in principle be solved.

Now we address the same problem in Einsteinian gravity. The energy-momentum tensor for N point particles of masses $m_1, ..., m_N$ is given by

$$T^{\mu\nu}(x) = \sum_k m_k(-g(x))^{-1/2}\delta^3(x - x_k(t))(dx_k^\mu(t)/dt)(dx_k^\nu/d\tau_k)$$

where τ_k is the proper time for the kth particle. It is given by

$$d\tau^2 = g_{\mu\nu}(x_k(t))dx_k^\mu(t)dx_k^\nu(t)$$

where

$$x_k^0(t) = t$$

is the universal coordinate time. The Einstein field equations corresponding to this energy-momentum tensor are

$$G_{\mu\nu} = R_{\mu\nu} - (1/2)Rg_{\mu\nu} = -KT_{\mu\nu}, K = 8\pi G$$

We find that

$$\int T^{\mu\nu}(t,r)\sqrt{-g(t,r)}f(t,r)dtd^3r = \sum T^{\mu\nu}(x)\sqrt{-g(x)}f(x)d^4x$$

$$= \sum_k m_k \int f(x_k(t))v_k^\mu(t)v_k^\nu(t)d\tau_k(t)$$

where

$$v_k^\mu(t) = dx_k^\mu/d\tau_k$$

is the four velocity of the kth particle. From this equation, we can infer by choosing different functions $f : \mathbb{R}^4 \to \mathbb{R}$, the particle trajectories as functions of coordinate time as well as their masses.

Application of the same ideas to Superstring theory: A superstring comprising of a Bosonic and a Fermionic part is given by

$$X^\mu(\tau,\sigma) = x^\mu + p^\mu\tau - i\sum_{n\neq 0}(\alpha^\mu(n)/n)exp(in(\tau-\sigma)) - i\sum_{n\neq 0}(\tilde{\alpha}^\mu(n)/n)exp(in(\tau+\sigma))$$

$$\psi^\mu(\tau,\sigma) = \psi_+(\tau,\sigma) + \psi_-(\tau,\sigma)$$

$$= \sum_n S_n^\mu exp(in(\tau-\sigma)) + \sum_n \tilde{S}_n^\mu exp(in(\tau+\sigma))$$

since these satisfy the string field equations

$$\partial_+\partial_- X^\mu = 0, \partial_-\psi_+^\mu = 0, \partial_+\psi_-^\mu = 0$$

where

$$\partial_+ = \partial_\tau + \partial_\sigma, \partial_- = \partial_\tau - \partial_\sigma$$

so that

$$\partial_+\partial_- = \partial_\tau^2 - \partial_\sigma^2$$

Note that the Lagrangian for the Bosonic part of the string is

$$L_B = (1/2)\partial_+ X^\mu.\partial_- X_\mu$$

while that of the Fermionic part is

$$L_F = -i\psi_+^T\partial_-\psi_- - i\psi_-^T\partial_+\psi_-$$

Note that $\psi_+^T\partial_-\psi_-$ is an abbreviation for $\psi_+^\mu\partial_-\psi_+\mu$ and likewise for the other term. If ψ_+ and ψ_- denote the canonical position fields for the Fermionic component of the superstring, then the corresponding canonical momenta are

$$\pi_+ = \partial L_F/\partial\partial_\tau\psi_+ = -i\psi_+,$$

$$\pi_- = \partial L_F.\partial_\tau\psi_- = -i\psi_-$$

so that the canonical anticommutation relations are

$$[\psi_+(\tau,\sigma),\psi_+(\tau,\sigma')]_+ = -\delta(\sigma-\sigma')$$

$$[\psi_-(\tau,\sigma),\psi_-(\tau,\sigma')]_+ = -delta(\sigma-\sigma')$$

These equations give

$$[S^\mu_n, S^\nu_m]_+ = \eta^{\mu\nu}\delta(n+m),$$

$$[\tilde{S}^\mu_n, \tilde{S}^\nu_m]_+ = \eta^{\mu\nu}\delta(n+m),$$

To obtain the Noether conserved currents for the Fermionic sector, we first observe that L_F is invariant under the infinitesimal transformations

$$\delta\psi_+ = \epsilon.\psi_-, \delta\psi_- = -\epsilon\psi_+$$

where ϵ is an infinitesimal parameter. The first conserved Noether current corresponding to this symmetry is then given by

$$J^- = (\partial L_F/\partial\partial_-\psi_+)\delta\psi_+ + (\partial L_F/\partial\partial_-\psi_-)\delta\psi_- = \psi_+^T\psi_+$$

which is obeys the conservation law

$$\partial_- J^- = 0$$

when the field equations are satisfied. Likewise, the second conserved current corresponding to this symmetry is

$$J^+ = (\partial L_F/\partial\partial_+\psi_-)\delta\psi_- = \psi_-^T\psi_-$$

which satisfies the conservation law

$$\partial_+ J^+ = 0$$

when the field equations are satisfied. Likewise, $J^+ = \psi_-^T\psi_-$ satisfies the conservation law

$$\partial_+ J^+ = 0$$

when the field equations are satisfied. The problem is can we calculate p^μ, the translational D-momentum of the string from measurements on the string observables ?

Exercise: Evaluate the Fourier series components of the energy-momentum tensor of a superstring. Also evaluate the Fourier series components of the supercurrent of a superstring. Show that the supercurrent field determine the Fermionic supersymmetry generators.

Acknowledgements:I am grateful to Prof.Hans Van Leunen, Prof.Andre Michaud and Prof.Steven Arthur L+angford for encouraging me to work on this problem and apply the method of determining all the charges and their locations from electromagnetic field measurements to Dirac's relativistic wave equation, by replacing the electronic charge which appears in this equation with functionals of the quantum electromagnetic field.

5.17 The quantum Boltzmann equation for a plasma

Suppose that the joint density matrix of N particles is $\rho(123...N)$. It satisfies the Schrodinger equation

$$i\partial_t\rho_t(12...N)=[\sum_{a=1}^N H_a+\sum_{1\leq a<b\leq N} V_{ab},\rho_t(12..N)]$$

In this equation, if we take a partial traced over $2,3,...,N$, we get

$$'i\partial_t\rho_{1t}=[H_1,\rho_{1t}]+(N-1)Tr_2[V_{12},\rho_{12}]$$

and if we take the trace of the same over $3,4,...,N$, we get

$$i\partial_t\rho_{12t}=[H_1+H_2+V_{12},\rho_{12t}]+(N-2)Tr_3[V_{13}+V_{23},\rho_{123t}]$$

We write

$$\rho_{123}=(1/3)(\rho_{12}\otimes\rho_3+\rho_{13}\otimes\rho_2+\rho_1\otimes\rho_{23})+g_{123}$$

where g_{123} is small. Then, neglecting second order of smallness terms like V multiplied with g_{123} gives us the approximate equation

$$i\partial_t\rho_{12t}=[H_1+H_2+V_{12},\rho_{12}]+((N-2)/3)Tr_3[V_{13}+V_{23},\rho_{12}\otimes\rho_3+\rho_{13}\otimes\rho_2+\rho_1\otimes\rho_{23}]$$

This is a bit hard to handle. So we content ourselves with the approximation

$$\rho_{12}=\rho_1\otimes\rho_1+g_{12}$$

where g_{12} is small. We then get approximately,

$$i\partial_t\rho_{1t}=[H_1,\rho_{1t}]+(N-1)Tr_2[V_{12},\rho_1\otimes\rho_1]$$

Writing

$$V_{12}=\sum_a W_{1a}\otimes W_{2a}$$

gives us

$$Tr_2[V_{12},\rho_1\otimes\rho_2]=\sum_a Tr(\rho_1 W_{2a})[W_{1a},\rho_1]$$

and the our Boltzmann equation becomes

$$i\partial_t\rho_1=[H_1,\rho_1]+(N-1)\sum_a Tr(\rho_1 W_{2a})[W_{1a},\rho_1]$$

Suppose we make the approximation

$$\rho_{123}=\rho_1\otimes\rho_1\otimes\rho_1+g_{123}$$

where g_{123} is small. Then we get

$$i\partial_t\rho_{12t}=[H_1+H_2+V_{12},\rho_{12}]+(N-2)Tr_3[V_{13}+V_{23},\rho_1\otimes\rho_1\otimes\rho_1]$$

Even this equation is hard to manipulate further without assuming some specific form of the interaction potential V_{12}. We consider

$$\rho_{12} = \rho_1 \otimes \rho_1 + g_{12},$$

$$\rho_{123} = (1/3)(\rho_{12} \otimes \rho_3 + \rho_{13} \otimes \rho_2 + \rho_1 \otimes \rho_{23}) + g_{123}$$

$$= \rho_1 \otimes \rho_1 \otimes \rho_1 + (1/3(g_{12} \otimes \rho_3 + g_{13} \otimes \rho_2 + \rho_1 \otimes g_{23}) + g_{123}$$

We first derive a differential equation for g_{12} after neglecting second order of smallness terms:

$$i\partial_t \rho_{12} = i\partial_t \rho_1 \otimes \rho_1 + i\rho_1 \otimes \partial_t \rho_1$$

$$+ i\partial_t g_{12}$$

$$= [H_1, \rho_1] \otimes \rho_1 + (N-1)Tr_2[V_{12}, \rho_1 \otimes \rho_1] \otimes \rho_1 - \rho_1 \otimes [H_1, \rho_1]$$

$$+(N-1)\rho_1 \otimes Tr_2[V_{12}, \rho_1 \otimes \rho_1] + i\partial_t g_{12}$$

$$= [H_1 + H_2 + V_{12}, \rho_{12}] + (N-2)Tr_3[V_{12} + V_{13}, \rho_{123}]$$

$$= [H_1 + H_2, \rho_1 \otimes \rho_1] + [V_{12}, \rho_1 \otimes \rho_1]$$

$$+(N-2)Tr_3[V_{13} + V_{23}, \rho_1 \otimes \rho_1 \otimes \rho_1]$$

After making the appropriate cancellations, we get

$$i\partial_t g_{12} =$$

$$= [V_{12}, \rho_1 \otimes \rho_1] + (N-2)Tr_3[V_{13} + V_{23}, \rho_1 \otimes \rho_1 \otimes \rho_1]$$

$$-(N-1)Tr_2[V_{12}, \rho_1 \otimes \rho_1] \otimes \rho_1$$

$$-(N-1)\rho_1 \times Tr_2[V_{12}, \rho_1 \otimes \rho_1]$$

Note that on writing

$$V_{12} = \sum_a W_{1a} \otimes W_{2a}$$

and using the fact that the $V_{jk}'s$ are identical copies of each other acting on different copies of the tensor product of two identical copies a Hilbert space just as the $H_k's$ are identical copies of each other acting on different copies of the same Hilbert space, we get

$$Tr_3[V_{13} + V_{23}, \rho_1 \otimes \rho_1 \otimes \rho_1]$$

$$= \sum_a [Tr(\rho_1 W_{2a})([W_{1a}, \rho_1] \otimes \rho_1 + \rho_1 \otimes [W_{1a}, \rho_1])]$$

A better approximation to the quantum Boltzmann equation can then be obtained by solving this equation for $g_{12}(t)$ and substituting it into the equation

$$i\partial_t \rho_1 = [H_1, \rho_1] + (N-1)Tr_2[V_{12}, \rho_{12}]$$

$$= [H_1, \rho_1] + (N-1)Tr_2[V_{12}, \rho_1 \otimes \rho_1 + g_{12}]$$

Formally this equation has the form

$$i\partial_t\rho_1(t) = [H_1, \rho_1(t)] + \delta.F_1(\rho_1(s), s \leq t)$$

where F is an operator valued nonlinear functional of $\rho_1(s), s \leq t$. This equation can be solved upto $O(\delta)$ using first order perturbation theory:

$$\rho_1(t) = U(t)\rho_1(0)U(t)^* + \delta.\int_0^t U(t-\tau)F_\tau(\rho_1(s), s \leq \tau)U(t-\tau)^* d\tau$$

where

$$U(t) = exp(-itH_1)$$

If we consider the Hamiltonian to comprise of an interaction between the particles and an electromagnetic field, then we can write

$$H_1 = (p_1+eA(t,r))^2/2m - e\Phi(t,r) \approx p_1^2/2m - e\Phi(t,r) + (e/2m)((p_1, A)+(A, p_1))$$

and the particle interaction potential as

$$V_{12} = V(|r_1 - r_2|)$$

In that case, in the position representation, we have

$$[p_1^2, \rho_1] = [p_1, \rho_1].p_1 + p_1.[p_1, \rho_1]$$

and noting that p_1 is represented by the kernel

$$p_1(r, r') = -i\nabla_r\delta^3(r-r') = i\nabla'_r\delta^3(r-r')$$

we get

$$[p_1, \rho_1](r, r') = -i\nabla_r\rho_1(r, r') + i\nabla'_r\rho_1(r, r')$$

$$[p_1, \rho_1].p_1(r, r') = \nabla_r.\nabla'_r\rho_1(r, r') - \nabla_r'^2\rho_1(r, r')$$

and likewise,

$$p_1.[p_1, \rho_1](r, r') = \nabla_r.\nabla'_r\rho_1(r, r') - \nabla_r^2\rho_1(r, r')$$

$$[(A, p_1), \rho_1](r, r') = [A_k p_{1k}, \rho_1](r, r') =$$

$$A_k(t, r)p_{1k}\rho_1(r, r') - \int \rho_1(r, r'')A_k(t, r'')p_{1k}(r'', r')dr''$$

$$= -i(A(t, r), \nabla_r)\rho_1(r, r') - i(\nabla'_r, A(t, r'))\rho_1(r, r'))$$

Therefore with neglect of nonlinear terms in the electromagnetic field, our position space representation of the quantum Boltzmann dynamics of the single particle density operator is given by

$$i\partial_t\rho_1(t, r, r') = (2m)^{-1}(2\nabla_r.\nabla'_r\rho_1(t, r, r') - (\nabla_r^2 + \nabla_r'^2)\rho_1(t, r, r'))$$

$$-(i/m)(A(t,r),\nabla_r)\rho_1(t,r,r') - i(\nabla'_r, A(t,r'))\rho_1(t,r,r')$$
$$-e\Phi(t,r)\rho_1(t,r,r') + e\Phi(t,r')\rho_1(t,r,r')$$
$$+nonlinearterms.$$

Assume that in principle, we have solved this equation for $\rho_1(t,r,r')$

Remark:

$$Tr_2[V_{12}\rho_1 \otimes \rho_1](r_1,r'_1) =$$
$$V(r_1,r_2)\delta(r_1 - r''_1)\delta(r_2 - r'_2)\rho_1(t,r''_1,r'_1)\rho_1(t,r'_2,r_2)d^3r'_2d^3r_2d$$
$$= \int V(r_1,r_2)\rho_1(t,r_1,r'_1)\rho_1(t,r_2,r_2)d^3r_2$$

and likewise,

$$Tr_2[(\rho_1 \otimes \rho_1)V_{12}](r_1,r'_1) =$$
$$\int \rho_1(t,r_1,r'_1)\rho_1(t,r_2,r_2)V(r'_1,r_2)d^3r_2$$

Combining these two equations, we get

$$Tr_2[V_{12},\rho_1 \otimes \rho_1](r,r') =$$
$$\int (V(r_1,r_2) - V(r'_1,r_2))\rho_1(t,r_1,r'_1)\rho_1(t,r_2,r_2)d^3r_2$$

which means that our quantum Boltzmann equation assumes the following form for the single particle density kernel evolution:

$$i\partial_t\rho_1(t,r,r') =$$
$$(2m)^{-1}(2\nabla_r.\nabla'_r\rho_1(t,r,r') - (\nabla_r^2 + \nabla_r'^2)\rho_1(t,r,r'))$$
$$-(i/m)(A(t,r),\nabla_r)\rho_1(t,r,r') - i(\nabla'_r, A(t,r'))\rho_1(t,r,r')$$
$$-e\Phi(t,r)\rho_1(t,r,r') + e\Phi(t,r')\rho_1(t,r,r')$$
$$+(N-1)\int (V(r,r_2) - V(r',r_2))\rho_1(t,r,r')\rho_1(t,r_2,r_2)d^3r_2$$

and we may replace $V(r,r')$ by $V(|r-r'|)$ in the case when the interaction potential between two particles is a only a function of the distance between them.

In principle, using perturbation theory, this quantum Boltzmann equation can be solved for to obtain the single particle density operator kernel $\rho_1(t,r,r')$ as a function of the electromagnetic field and then the average dipole moment of the electron in this electromagnetic field will be given by

$$\mathbf{p}(t) = \int (-e\mathbf{r})\rho_1(t,\mathbf{r},\mathbf{r})d^3r$$

and its average magnetic moment by

$$\mathbf{m}(t) = \int (-e\mathbf{L}(r,r')/2m)\rho_1(t,\mathbf{r}',\mathbf{r})d^3rd^3r'$$

where $\mathbf{L}(r,r')$ is the kernel of the orbital angular momentum operator in the position representation. This would then solve the problem of explaining the origin of permittivity and permeability of the plasma from the quantum statistical mechanical point of view. The angular momentum kernel is obtained as follows:

$$L_x(r,r') = (yp_z - zp_y)(r,r') = -iy\delta'(z-z')\delta(x-x')\delta(y-y') + iz\delta(x-x')\delta'(y-y')\delta(z-z')$$

$$L_y(r,r') = (zp_x - xp_z)(r,r') = -iz\delta'(x-x')\delta(y-y')\delta(z-z') + ix\delta(x-x')\delta(y-y')\delta'(z-z')$$

$$L_z(r,r') = (xp_y - yp_x)(r,r') = -ix\delta(x-x')\delta'(y-y')\delta(z-z') + iy\delta'(x-x')\delta(y-y')\delta(z-z')$$

where

$$r = (x,y,z), r' = (x',y',z')$$

Then the components of the average magnetic moment can be expressed as

$$m_x(t) = \int (-eL_x/2m)(r,r')\rho_1(t,r',r)d^3r'd^3r$$

$$= (ie/2m)\int (y\delta'(z-z')\delta(x-x')\delta(y-y') - z\delta(x-x')\delta'(y-y')\delta(z-z'))\rho_1(t,r',r)d^3r'd^3r$$

$$= (ie/2m)[\int [-y\partial_z\rho_1(t,r',r)|_{r'=r} + z\partial_y\rho_1(t,r',r)|_{r'=r}]dxdydz$$

and likewise for the other components. We could arrange these calculations in vectorial notation:

$$\mathbf{L}(r,r') = -i(r\times\nabla)(r,r') = -ir\times\nabla_r\delta(r-r')$$

and hence

$$\int \mathbf{L}(r,r')\rho_1(t,r',r)d^3rd^3r' = -i\int r\times(\nabla_r\delta(r-r'))\rho_1(t,r',r)d^3rd^3r'$$

$$= -i\int r\times\nabla_1\rho_1(t,r,r)d^3r$$

where ∇_1 stands for the gradient w.r.t the first argument:

$$\nabla_1\rho_1(t,r,r) = (\nabla_{r_1}\rho_1(t,r_1,r))|_{r_1=r}$$

Perturbative solution of the Boltzmann equation :

$$i\partial_t\rho_1(t) = [H_1,\rho_1(t)] + \delta(N-1)Tr_2[V_{12},\rho_1(t)\otimes\rho_1(t)]$$

ie, $B(.)$ is a Gaussian process with a.s. continuous sample paths having zero mean and covariance $min(t,s)$, or in other words, $B(.)$ is a Brownian motion process over $[0,1]$.

Remark: To prove that $B(.)$ is a Gaussian process, it suffices to show that if $\{t_1,...,t_k\}$ is a finite set of points in $[0,1]$, then $(B(t_1),...,B(t_k))$ is a Gaussian random vector. But this is immediately a consequence of the fact that $(B_N(t_1),...,B_N(t_k))$ is a Gaussian random vector which converges in distribution since it converges in probability since it converges a.s since the process $B_N(.)$ converges uniformly a.s.

[2] The law of the iterated logarithm: This result states that if $B(.)$ is standard BM, then

$$limsup_{t\to\infty}\frac{B(t)}{\sqrt{2t.loglog(t)}}=1a.s$$

Equivalently, since $tB(1/t)$ is also a BM, we can state the law of the iterated logarithm as

$$limsup_{t\to 0}\frac{B(t)}{2t.loglog(1/t)}=1a.s$$

Equivalently since $-B(t)$ is also a BM, this law can also be stated as

$$liminf_{t\to\infty}\frac{B(t)}{2t.loglog(t)}=-1$$

Intuitively what these result state is that for very large times t, $B(t)$ almost surely oscillates between the two bounding curves $x=\pm\sqrt{2t.loglog(t)}$.

To prove this result, define

$$h(t)=\sqrt{2t.lnln(1/t)}, 0<t<1$$

so that for $0<\theta<1$ and $n=0,1,...$, we have

$$h(\theta^n)=\sqrt{2\theta^n.ln(n.ln(1/\theta)}$$

Then we use Doob's Martingale's inequality in the form

$$P(max_{0<s<t}(B(s)-\lambda s/2)>\beta)=P(max_{0<s<t}exp(\lambda.B(s)-\lambda^2 s/2)\leq exp(-\lambda\beta))$$

$$\leq exp(-\lambda\beta), \lambda>0$$

In this inequality, we choose

$$\beta=h(\theta^n)/2, \lambda=(1+\delta)\theta^{-n}h(\theta^n)$$

to get

$$P(max_{\theta^{n+1}<s<\theta^n}(B(s)-(1+\delta)\theta^{-n}h(\theta^n)s/2)>h(\theta^n))\leq exp(-(1+\delta)\theta^{-n}h(\theta^n)^2/2)$$

$$= \alpha_k(-i\partial_k\rho(t,r,r')) - i\partial'_k\rho(t,r,r')\alpha_k$$

and

$$[\beta,\rho](r,r') = \beta.\rho(t,r,r') - \rho(t,r,r')\beta$$

Further,

$$((\alpha,A)\rho)(r,r') = A_k(t,r)\alpha_k\rho(t,r,r') = (\alpha, A(t,r))\rho(t,r,r')$$

$$(\rho(\alpha,A))(r,r') = \rho(t,r,r')(\alpha, A(t,r'))$$

Thus,

$$[(\alpha,A),\rho](r,r') = (\alpha,A(t,r))\rho(t,r,r') - \rho(t,r,r')(\alpha,A(t,r'))$$

and likewise,

$$[\Phi,\rho](r,r') = (\Phi(t,r) - \Phi(t,r'))\rho(t,r,r')$$

5.18 Some other remarks on Lie algebras

Let $\mathfrak{g}$ be a complex SSLA(Semisimple Lie algebra) and let $\mathfrak{h}$ be a CSA (Cartan subalgebra). Consider the root space decompostion

$$\mathfrak{g} = \mathfrak{h} \oplus \bigoplus_{\alpha\in\Delta} \mathfrak{g}_\alpha$$

Note that Δ is a finite subset of $\mathfrak{h}^*$. We have that

$$[H,X] = \alpha(H)X, H \in \mathfrak{h}, X \in \mathfrak{g}_\alpha$$

$$[H,H'] = 0, H, H' \in \mathfrak{h}$$

and $ad(H), H \in \mathfrak{h}$ are semisimple operators on $\mathfrak{g}$ which is what makes the above root space decomposition possible. It is clear that

$$[\mathfrak{g}_\alpha, \mathfrak{g}_\beta] \subset \mathfrak{g}_{\alpha+\beta}, \alpha, \beta \in \Delta$$

where $\mathfrak{g}_{\alpha+\beta}$ is to be taken as zero if $\alpha+\beta \neq 0$ and $\alpha+\beta$ is not a root and as $\mathfrak{h}$ if $\alpha+\beta = 0$.

This is because if $X \in \mathfrak{g}_\alpha, Y \in \mathfrak{g}_\beta$, then by the Jacobi identity,

$$[H,[X,Y]] = -([X,[Y,H]] + [Y,[H,X]] = (\alpha+\beta)(H)[X,Y], H \in \mathfrak{h}$$

Further,

$$[\mathfrak{h}, \mathfrak{g}_\alpha] \subset \mathfrak{g}_\alpha$$

as again follows by use of the Jacobi identity and the Abelian character of $\mathfrak{h}$.

Note that if $X \in \mathfrak{g}_\alpha, Y \in \mathfrak{g}_{-\alpha}$, then for any $H \in \mathfrak{h}$, we have by the Jacobi identity that

$$[H,[X,Y]] = 0$$

and hence since $\mathfrak{h}$ is maximal Abelian, it follows that

$$[X,Y] \in \mathfrak{h}$$

In other words,

$$[\mathfrak{g}_\alpha, \mathfrak{g}_{-\alpha}] \subset \mathfrak{h}, \alpha \in \Delta$$

It is clear that $< .,. >$ is non-singular on $\mathfrak{h} \times \mathfrak{h}$ as well as on $\mathfrak{g}_\alpha \times \mathfrak{g}_{-\alpha}$ for the following reasons:

$$< X, [H,Y] >=< [X,H], Y >= \beta(H) < X,Y >= -\alpha(H) < X,Y >, X \in \mathfrak{g}_\alpha, Y \in \mathfrak{g}_\beta, H \in \mathfrak{h}$$

If $\beta+\alpha \neq 0$, it follows that there exists an $H \in \mathfrak{h}$ for which $(\beta+\alpha)(H) \neq 0$ and therefore $< X,Y >= 0$. In other words, we have proved that if $\alpha, \beta \in \Delta$ and $\beta \neq -\alpha$, then $\mathfrak{g}_\beta \perp \mathfrak{g}_\alpha$. Further $\mathfrak{h} \perp \mathfrak{g}_\alpha \forall \alpha \in \Delta$. This follows from the identity

$$\alpha(H') < H,X >=< H,[H',X] >=< [H,H'],X >= 0, H,H' \in \mathfrak{h}, X \in \mathfrak{g}_\alpha$$

which implies the $< H,X >= 0, H \in \mathfrak{h}$ since $\alpha \neq 0$ in $\mathfrak{h}^*$. Since therefore, for a given $\alpha \in \Delta$, $\mathfrak{g}_\alpha$ is orthogonal to $\mathfrak{h}$ as well as to $\mathfrak{g}_\beta$ for all roots $\beta \neq -\alpha$, the root space decomposition implies that for any $\alpha \in \Delta$, $X \in \mathfrak{g}_\alpha$ cannot be orthogonal to $\mathfrak{g}_{-\alpha}$ for other wise, it would be orthogonal to $\mathfrak{g}$ which is false since for a SSLA, the Cartan-Killing form is non-singular. Likewise since $\mathfrak{h}$ is orthogonal to $\mathfrak{g}_\alpha \forall \alpha \in \Delta$, it follows from the root space decomposition and the non-singularity of the Cartan-Killing form that $\mathfrak{h}$ cannot be orthogonal to itself, ie, the Cartan-Killing form is non-singular on $\mathfrak{h} \times \mathfrak{h}$.

Remark:$\Delta = -\Delta$, ie, if $\alpha \in \Delta$, then $-\alpha \in \Delta$. For suppose that $\alpha \in \Delta$ but $-\alpha \notin \Delta$. Then $\mathfrak{g}_\alpha$ is orthogonal to $\mathfrak{h}$ as well to all the $\mathfrak{g}_\beta, \beta \in \Delta$ since $-\alpha \notin \Delta$. Thus by the root space decomposition, $\mathfrak{g}_\alpha$ is orthogonal to $\mathfrak{g}$ which contradicts the non-degeneracy of the Cartan-Killing form on $\mathfrak{g}$.

Now we show that $dim\mathfrak{g}_\alpha = 1 \forall \alpha \in \Delta$. In fact, since the Cartan-Killing form is non-singular on $\mathfrak{g}_\alpha \times \mathfrak{g}_{-\alpha}$, it follows that we can select $X_\alpha \in \mathfrak{g}_\alpha$ and $X_{-\alpha} \in \mathfrak{g}_{-\alpha}$ so that

$$< X_\alpha, X_{-\alpha} > \neq 0$$

and hence, we can define $H_\alpha \in \mathfrak{h}$ so that

$$[X_\alpha, X_{-\alpha}] =< X_\alpha, X_{-\alpha} > H_\alpha, \alpha \in \Delta$$

Then, it is clear that

$$< H_\alpha, H >= \frac{< [X_\alpha, X_{-\alpha}], H >}{< X_\alpha, X_{-\alpha} >}$$

$$= \frac{< X_\alpha, [X_{-\alpha}, H] >}{< X_\alpha, X_{-\alpha} >}$$

$$= \alpha(H), H \in \mathfrak{h}$$

Now choose complex numbers $c_\alpha, c_{-\alpha}$ such that

$$c_\alpha c_{-\alpha} < X_\alpha, X_{-\alpha} >= 2/\alpha(H_\alpha)$$

and define

$$\bar{X}_\alpha = c_\alpha X_\alpha, \bar{X}_{-\alpha} = c_{-\alpha} X_{-\alpha}$$

Also define

$$\bar{H}_\alpha = 2H_\alpha/\alpha(H_\alpha)$$

(We shall soon be showing that $\alpha(H_\alpha) \neq 0$ so that division by it is justified).

Then, it is easy to see that

$$[\bar{H}_\alpha, \bar{X}_\alpha] = 2\bar{X}_\alpha, [\bar{H}_\alpha, X_{-\alpha}] = -2\bar{X}_{-\alpha},$$

$$[\bar{X}_\alpha, \bar{X}_{-\alpha}] = \bar{H}_\alpha$$

or equivalently, $\{\bar{H}_\alpha, \bar{X}_\alpha, \bar{X}_{-\alpha}\}$ forms a standard basis for an $\mathfrak{sl}(2,\mathbb{C})$ Lie algebra. Hence, we shall denote $\bar{X}_\alpha$ by X_α and $\bar{X}_{-\alpha}$ by $X_{-\alpha}$ for notational convenience. Thus, in our new notation, $\{\bar{H}_\alpha, X_\alpha, X_{-\alpha}\}$ forms a standard $\mathfrak{sl}(2,\mathbb{C})$ triplet.

Result: $span\{H_\alpha : \alpha \in \Delta\} = \mathfrak{h}^*$. To see this, it suffices to show that (in view of the non-singularity of the Cartan-Killing form on $\mathfrak{h} \times \mathfrak{h}$ that if $H \in \mathfrak{h}$ is such that $\alpha(H) =< H_\alpha, H >= 0 \forall \alpha \in \Delta$, then $H = 0$. But from the definition of the Cartan-Killing form,

$$< H', H >= \sum_{\beta \in \Delta} dim(\mathfrak{g}_\beta)\beta(H')\beta(H) = 0 \forall H' \in \mathfrak{h}$$

since by hypothesis, $\beta(H) = 0 \forall \beta \in \Delta$. But then since $< .,. >$ is non-singular on $\mathfrak{h} \times \mathfrak{h}$, it follows that $H = 0$ and we are done.

Result: For any $\alpha, \beta \in \Delta$, there exists a real rational number $q_{\beta,\alpha}$ such that

$$\beta(H_\alpha) = q_{\beta,\alpha}\alpha(H_\alpha)$$

To see this, we first construct a maximal chain of roots $\beta + k\alpha, k = -p, -p + 1, ..., q - 1, q$ where p, q are non-negative integers. By maximal chain, we mean that $\beta - (p+1)\alpha$ and $\beta + (q+1)\alpha$ are not roots. Then, it follows that

$$ad(X_{-\alpha})(\mathfrak{g}_{\beta-p\alpha}) = 0,$$

$$ad(X_\alpha)(\mathfrak{g}_{\beta+q\alpha}) = 0,$$

$$ad(X_{-\alpha})(\mathfrak{g}_{\beta+k\alpha}) \subset \mathfrak{g}_{\beta+(k-1)\alpha}$$

for $-p < k \leq q$,

$$ad(X_\alpha)(\mathfrak{g}_{\beta+k\alpha}) \subset \mathfrak{g}_{\beta+(k+1)\alpha}$$

for $-p \leq k < q$ and

$$ad(\bar{H}_\alpha)(\mathfrak{g}_{\beta+k\alpha}) \subset \mathfrak{g}_{\beta+k\alpha}$$

for $-p \leq k \leq q$. In particular, we see that the vector space

$$V_{\beta,\alpha} = \bigoplus_{k=-p}^{q} \mathfrak{g}_{\beta+k\alpha}$$

is invariant under the $sl(2,\mathbb{C})$ adjoint algebra $\{ad(\bar{H}_\alpha), ad(X_\alpha), ad(X_{-\alpha})\}$. Thus,

$$Tr(ad(\bar{H}_\alpha)|_{V_{\beta,\alpha}} = Tr([ad(X_\alpha), ad(X_{-\alpha})]|_{V_{\beta,\alpha}}) = 0$$

Evaluating this trace gives us

$$\sum_{k=-p}^{q} dim(\mathfrak{g}_{\beta+k\alpha})(\beta+k\alpha)(\bar{H}_\alpha) = 0$$

or equivalently since H_α is a scalar times $\bar{H}_\alpha$, we get

$$\beta(H_\alpha).(\sum_{k=-p}^{q} dim(\mathfrak{g}_{\beta+k\alpha}) + (\sum_{k=-p}^{q} kdim(\mathfrak{g}_{\beta+k\alpha}))\alpha(H_\alpha) = 0$$

which completes the proof once we note that

$$q_{\beta,\alpha} = -\frac{\sum_{k=-p}^{q} k.dim(\mathfrak{g}_{\beta+k\alpha})}{\sum_{k=-p}^{q} dim(\mathfrak{g}_{\beta+k\alpha}}$$

is a real rational number.

Result:$\alpha(H_\alpha) \neq 0$ for any $\alpha \in \Delta$. To see this, suppose that for some $\alpha \in \Delta$, we have that $\alpha(H_\alpha) = 0$. Then by the previous result, $\beta(H_\alpha) = q_{\beta,\alpha}\alpha(H_\alpha) = 0 \forall \beta \in \Delta$. This implies that $H_\alpha = 0$ since as already noted above span $\{\Delta\} = \mathfrak{h}^*$. Thus, $\alpha = 0$ since $\alpha(H) =< H_\alpha, H >, H \in \mathfrak{h}$. This contradiction proves the result.

Result:For any $\alpha \in \Delta$, $\alpha(H_\alpha)$ is a real, positive rational number. To prove this, we use the two results proved above, namely that $\alpha(H_\alpha)$ is a nonzero complex number and that $\beta(H_\alpha) = q_{\beta,\alpha}\alpha(H_\alpha)$ for any $\beta \in \Delta$ where $q_{\beta,\alpha}$ is a real number (in fact a real rational number). Then, we get

$$\alpha(H_\alpha) =< H_\alpha, H_\alpha >= \sum_{\beta\in\Delta} \beta(H_\alpha)^2 dim(\mathfrak{g}_\beta)$$

$$= \sum_{\beta\in\Delta} dim(\mathfrak{g}_\beta) q_{\beta,\alpha}^2 \alpha(H_\alpha)^2$$

(Note that $dim\mathfrak{g}_\beta \geq 1 \forall \beta \in \Delta$). Since $\alpha(H_\alpha) \neq 0$, we can cancel it from both the sides to get

$$\alpha(H_\alpha) = [\sum_{\beta\in\Delta} dim(\mathfrak{g}_\beta) q_{\beta,\alpha})^2]^{-1}$$

which proves the claim.

Result: $dim\mathfrak{g}_\alpha = 1 \ \forall \alpha \in \Delta$.
Let $\alpha \in \Delta$ and define the subspace

$$V = span\{X_{-\alpha}\} \oplus \mathfrak{h} \oplus \mathfrak{g}_\alpha \oplus \mathfrak{g}_{2\alpha} \oplus ... \oplus \mathfrak{g}_{m\alpha} \oplus ..$$

It is clear that this series terminates after a finite number of steps because $\mathfrak{g}$ is finite dimensional. Further, it is clear that V is invariant under $\{ad(X_{-\alpha}), ad(X_\alpha), ad(\bar{H}_\alpha)\}$ and therefore since

$$ad(\bar{H}_\alpha) = [ad(X_\alpha), ad(X_{-\alpha})]$$

it follows that

$$Tr(ad(\bar{H}_\alpha)|_V) = 0$$

5.19 Question Paper on Matrix Theory

Attempt any four questions. Each question carries five marks.

[1] Let $\mathbf{A}$ be an $m \times n$ matrix with $m > n$ and having rank n. Write down the general structure of the singular value decomposition of $\mathbf{A}$. Explain how using this SVD, you will obtain the least squares solution to the problem of calculating $\theta \in \mathbb{R}^n$ so that

$$(\mathbf{x} - \mathbf{A}\theta)^T(\mathbf{x} - \mathbf{A}\theta)$$

is minimum for a given $\mathbf{x} \in \mathbb{R}^n$. How will you modify your method if we have to solve the weighted least squares problem of minimizing

$$(\mathbf{x} - \mathbf{A}\theta)^T\mathbf{W}(\mathbf{x} - \mathbf{A}\theta)$$

where $\mathbf{W}$ is a positive definite $n \times n$ matrix.

[2] Write down the root space decomposition of the Lie algebra $\mathfrak{sl}(n, \mathbb{C})$, ie, the Lie algebra of all $n \times n$ complex matrices having zero trace. Identify clearly the Cartan subalgebra and the root vectors. Prove that this Lie algebra is indeed semisimple by calculating it Cartan-Killing form and showing that this form is non-singular.

[3] Let $\mathbf{A}$ be an $n \times n$ matrix having exactly two Jordan blocks of sizes $n_1 \times n_1$ with eigenvalue c_1 and $n_2 \times n_2$ with eigenvalue c_2 where $c_1 \neq c_2$. Write down the explicit forms of $exp(t\mathbf{A})$ and $(\lambda I - \mathbf{A})^{-1}$ in terms of these Jordan blocks and hence deduce that

$$\int_0^\infty exp(t\mathbf{A})exp(-\lambda t)dt = (\lambda I - \mathbf{A})^{-1}$$

provided that $Re(\lambda) > Re(c_k), k = 1, 2$.

[4] Using the primary decomposition theorem, prove that if T, S are complex $n \times n$ matrices such that $ad(T)^m(S) = 0$ for some positive integer m, then S leaves both the subspaces $\bigcup_{k\geq 1} \mathcal{N}(ad(T)^k)$ and $\bigcap_{k\geq 1} \mathcal{R}(ad(T)^k)$ invariant.

[5] Write short notes on the following:
[a] The primary decomposition theorem.
[b] Construction of a Cartan subalgebra of a semisimple Lie algebra.
[c] LDU and UDL decomposition of a positive definite matrix with applications to signal prediction theory.
[d] Gram-Schmidt orthonormalization with application to the QR decomposition of a full column rank matrix.

5.20 Study project on quantum antennas

Consider a cavity resonator B with boundary ∂B inside which we have confined both an electromagnetic field with vector potential expansion

$$\mathbf{A}(t, \mathbf{r}) = \sum_n a(t, n)\mathbf{u}_n(\mathbf{r})$$

and a corresponding Dirac field expansion

$$\psi(t, \mathbf{r}) = \sum_n \mathbf{c}(t, n)v_n(\mathbf{r})$$

where the 3-vector valued basis functions $\mathbf{u}_n(\mathbf{r})$ are chosen so that the corresponding electric and magnetic fields satisfy the appropriate boundary conditions and the scalar basis functions $v_n(\mathbf{r})$ also satisfy the appropriate boundary conditions. Owing to the Hermitianity of the Laplacian and the Dirac Hamiltonian, these basis functions can be chosen to be orthonormal:

$$< \mathbf{u}_n, \mathbf{u}_m >= \delta_{nm}, < v_n, v_m >= \delta_{nm}.$$

Since $divE = 0$ because there is no volume charge density and $E = -\nabla\Phi - \partial_t A$, if we adopt the Coulomb gauge, $\nabla^2\Phi = 0, divA = 0$ and hence $\Phi = 0$. Thus in the Coulomb gauge $E = -\partial_t A$ and $B = curlA$. The coulomb gauge condition $divE = 0$ implies $div\mathbf{u}_n = 0$. The total energy density in the electromagnetic field within the cavity is then

$$U_F = (1/2) \int_B [(\partial_t A)^2 + (curlA)^2]d^3x$$

and by using the orthonormality of the basis functions, we get

$$U_F = (1/2) \sum_n (\partial_t a(t, n))^2 + (1/2) \sum_n a(t, n)^2 \int_B (curl\mathbf{u}_n(\mathbf{r}))^2 d^3x$$

Remark: Within the cavity,

$$(\nabla^2 + \omega(n)^2)\mathbf{u}_n(\mathbf{r}) = 0$$

where the $\omega(n)'s$ are the characteristic frequencies of oscillation. Then

$$\int_B (curl\mathbf{u}_n(r), curl\mathbf{u}_m(r))d^3x$$

$$= -\int (\mathbf{u}_n, \nabla^2\mathbf{u}_m)(r)d^3x$$

$$= \omega(n)^2\delta_{n,m}$$

Thus,

$$U_F = (1/2)\sum_n((\partial_t a(t,n))^2 + \omega(n)^2 a(t,n)^2)$$

Likewise, we can quantize the Dirac field and its energy (Hamiltonian) within the cavity is given by

$$U_D = \int_B \psi(t,r)^*(\alpha, -i\nabla) + \beta m)\psi(t,r)d^3x$$

Now, the Dirac equation gives

$$\sum_n i\partial_t\mathbf{c}(t,n)v_n(r) = H_D\sum_n \mathbf{c}(t,n)v_n(r)$$

where

$$H_D = (\alpha, -i\nabla) + \beta m$$

This gives

$$i\partial_t c_l(t,n) = \sum_{m,k} < v_n(r), H_{Dlk}v_m > c_k(t,m)$$

and then, the second quantized Dirac Hamiltonian is given by

$$U_D = \sum_{lknm} c_l(t,n)^* c_k(t,m) < v_n, H_{Dlk}v_m >$$

In order to interpret this formula in terms of electron-positron creation and annihilation operators, we require to first diagonalize the quadratic form U_D.

5.21 Heat and mass transfer equations in a fluid

Study project: Write down the Navier-Stokes equation, the equation of continuity and the energy equation for an adiabatic fluid.

5.22 Quantum electrodynamics in a background medium described by a permittivity and permeability function

We have to quantize the Maxwell equations

$$[\epsilon(\mu\nu\alpha\beta, x) * F^{\alpha\beta}(x)]_{,\nu} = 0$$

where $*$ denotes space-time convolution and

$$F_{\mu\nu}(x) = A_{\nu,\mu}(x) - A_{\mu,\nu}(x)$$

The Lagrangian for density for the electromagnetic field is

$$L = (-1/4)F_{\mu\nu}(x).(\epsilon(\mu\nu\rho\sigma, x) * F^{\rho\sigma}(x))$$

The above Maxwell equations can be expressed in the space-time frequency domain as

$$\hat{\epsilon}(\mu\nu\alpha\beta, k)k_\nu \hat{F}^{\alpha\beta}(k) = 0$$

where

$$\hat{F}^{\mu\nu}(k) = \int F^{\mu\nu}(x)exp(-ik.x)d^4x$$

$$\hat{\epsilon}(\mu\nu\alpha\beta, k) = \int \epsilon(\mu\nu\alpha\beta, x)exp(-ik.x)d^4x$$

where

$$k.x = k^0x^0 - \sum_{r=1}^{3} k^r x^r, x^0 = t, (x^r)_{r=1}^3 = \mathbf{r}$$

Further,

$$\hat{F}_{\mu\nu}(k) = k_\mu \hat{A}_\nu(k) - k_\nu \hat{A}_\mu(k)$$

and so our Maxwell equations assume the form

$$\hat{\epsilon}(\mu\nu\alpha\beta, k)k_\nu(k^\alpha \hat{A}^\beta(k) - k^\beta \hat{A}^\alpha(k)) = 0$$

We can adopt the gauge condition

$$\hat{\epsilon}(\mu\nu\alpha\beta, k)k_\nu \hat{A}^\beta(k) = 0$$

and then we get the generalized wave equation in the four wave vector domain

$$\hat{\epsilon}(\mu\nu\alpha\beta, k)k_\nu k^\beta \hat{A}^\alpha(k) = 0 - - - (1)$$

from which the dispersion relation can easily be obtained

$$det((\hat{\epsilon}(\mu\nu\alpha\beta, k)k_\nu k^\beta))_{0\le\mu,\alpha\le 3} = 0 - - - (2)$$

Note that this equation is an eighth degree polynomial equation in the k^μ and hence it will generally have eight solutions for k^0 in terms of $(k^r)_{r=1}^3$. In the special case when the medium is the vacuum, we have

$$\hat{\epsilon}(\mu\nu\alpha\beta, k) = \delta_{\mu\alpha}.\delta_{\nu\beta}$$

and the dispersion relation reduces to the standard one

$$k_\nu k^\nu = 0$$

Assume that the solution to the above dispersion relation (1) can be expressed as

$$k^0 = \omega_m(K), K = (k^r)_{r=1}^3, m = 1, 2, ..., 8$$

Also assume that a set of linearly independent eigenvectors $\hat{A}^\mu(k)$ that satisfy both the gauge condition and the dispersion relation are given by

$$e^\mu(K, s), s = 1, 2, ..., 8$$

Then it is clear that the electromagnetic four potential can be expanded as

$$A^\mu(x) = \int [a(K, s)e^\mu(K, s)exp(-ik^{(s)}.x)]d^3K$$

where the sum is over $s = 1, 2, ..., 8$ and

$$k^{(s)} = (\omega_s(K), K), K = (k^r)_{r=1}^3$$

More precisely, we can write

$$A^\mu(x) = \int [a(K, s)e^\mu(K, s)exp(-i(\omega_s(K)t - K.r))d^3K$$

In order that this be a real field we must assume that for each s,

$$a(K, s)^* e^\mu(K, s)^* = a(-K, s')e^\mu(-K, s')$$

for some s' and that for this s',

$$-\omega_s(K) = \omega_{s'}(-K)$$

The energy of this electromagnetic field can be obtained in the frequency domain as follows. First, the Lagrangian density in the four wave vector domain is

$$\hat{L} = (-1/4)\epsilon(\mu\nu\alpha\beta, k)\hat{F}_{\mu\nu}(k)\hat{F}^{\alpha\beta}(k)$$

Then the position fields in the four wave vector domain being $\hat{A}_\mu(k)$, it follows that the corresponding momentum fields in the four wave vector domain are

$$\hat{\pi}^\nu(k) = \partial\hat{L}/\partial(k_0\hat{A}_\nu(k))$$

$$= \epsilon(0\nu\alpha\beta, k)\hat{F}^{\alpha\beta}(k)$$

and hence the Hamiltonian density in the four vector domain is given by applying the wave vector domain Legendre transformation to the Lagrangian density:

$$\mathcal{H}(k) = \hat{\pi}^{\nu}(k)k_0\hat{A}_{\nu}(k) - \hat{L}(k)$$

$$= \epsilon(0\nu\alpha\beta, k)\hat{F}^{\alpha\beta}(k)k_0\hat{A}_{\nu}(k) + (1/4)\hat{\epsilon}(\mu\nu\alpha\beta, k)\hat{F}_{\mu\nu}(k)\hat{F}^{\alpha\beta}(k)$$

From these calculations, it is clear that the total field Hamiltonian, ie field energy can be expressed as

$$U = \int C_1(\mu\nu\alpha\beta, k)\hat{F}_{\mu\nu}(k)^*\hat{F}_{\alpha\beta}(k)d^3K$$

here it is understood thqat we substitute for k^0, its dispersion relation values in terms of $K = (k^r)_{r=1}^3$ and then sum up over all the roots. Here, $C_1(\mu\nu\alpha\beta, k)$ is a function of the permittivity-permeability tensor $\epsilon(\mu\nu\alpha\beta, k)$ (not a functional, simply an ordinary function). Alternately, substituting for $\hat{F}_{\mu\nu}(k)$ its value in term of $\hat{A}_{\mu}(k)$, we can express the field energy in the form

$$U = \int C(K, s)a(K, s)^*a(K, s)d^3K$$

where the function $C(K, s)$ is derived from $C_1(\mu\nu\alpha\beta, k)$ and the polarization vectors $e^{\mu}(K, s)$. Note that

$$\hat{F}_{\mu\nu}(k) = k_{\mu}\hat{A}_{\nu}(k) - k_{\nu}\hat{A}_{\mu}(k)$$

which in turn equals

$$a(K, s)(k_{\mu}e_{\nu}(K, s) - k_{\nu}e_{\mu}(K, s))\delta(k^0 - \omega_s(K))$$

the sum over s being understood.

Models for the refractive index of a material based on classical and quantum physics

[a] Classical physics:

Consider an electron of effective mass $m(r)$ and charge $e(r)$ moving around its equilibrium position w.r.t the atomic nucleus. Let $\gamma(r)$ denote the damping coefficient during the electron's motion and let $E(t, r)$ denote the applied external electric field. If $\xi(t)$ is the displacement of the electron relative to the nucleus, then we have from classical Newtonian mechanics,

$$m(r + \xi(t))\xi''(t) + \gamma(r + \xi(t))\xi'(t) + K(r + \xi(t))\xi(t) = -eE(t, r + \xi(t))/m$$

and since $\xi(t)$ is very small, we can assume that

$$K(r + \xi) \approx K(r), \gamma(r + \xi) \approx \gamma(r), m(r + \xi) \approx m(r)$$

Then if the electric field has frequency ω, we can write

$$E(t, r) = Re(E_0(r)exp(i\omega t))$$

and writing

$$\xi(t) = Re(\xi_0 exp(i\omega t))$$

we find on substituting into the equation of motion

$$[m(r)(\omega_0(r)^2 - \omega^2) + i\gamma(r)\omega]\xi_0 = -eE_0(r), (K(r)/m(r))^{1/2} = \omega_0(r)$$

so the dipole moment of the electron can be expressed in the form

$$p(t, r) = -e\xi(t) = Re(p_0(r)exp(i\omega t))$$

where

$$p_0(r) = e^2 E_0(r)/[m(r)(\omega_0(r)^2 - \omega^2) + i\gamma(r)\omega]$$

If there are $N(r)$ atoms per unit volume at r, then the polarization phasor (ie dipole moment per unit volume) is given by

$$P(\omega, r) = N(r)e^2 E_0(r)/[m(r)(\omega_0(r)^2 - \omega^2) + i\gamma(r)\omega]$$

from which, we deduce that the complex refractive index of the material as a function of the location is given by

$$n(\omega, r) = (1 + Ne^2\epsilon_0^{-1}/[m(r)(\omega_0(r)^2 - \omega^2) + i\gamma(r)\omega])^{1/2}$$

$$\approx 1 + N(r)e^2\epsilon_0^{-1}/2[m(r)(\omega_0(r)^2 - \omega^2) + i\gamma(r)\omega]$$

for low electron densities $N(r)$

Remark: The complex permittivity is given by

$$\epsilon(\omega, r) = \epsilon_0(1 + n(r))$$

so that if we separate out the real and imaginary parts as

$$\epsilon(\omega, r) = \epsilon_R(\omega, r) + i\epsilon_I(\omega, r)$$

then the true permittivity of the medium is given by $\epsilon_R(\omega, r)$ while the true conductivity is given by

$$\sigma(\omega, r) = -\omega\epsilon_I(\omega, r)$$

In case, the medium in which the electron moves is anisotropic, $m(r), \gamma(r), K(r)$ become 3×3 matrices and we get the result that the permittivity becomes a 3×3 matrix given by

$$\epsilon(\omega, r) = \epsilon_0(I_3 + N(r)e^2((K(r) - m(r)\omega^2) + i\gamma(r)\omega)^{-1})$$

This is independent of the temperature and also of the electric. However, it depends on the frequency and hence the wavelength of electromagnetic radiation. Further, this formula can be generalized to the situation in which there are $N(\omega_0, r)d\omega_0$ atoms per unit volume whose electrons are bound to them by spring constants having natural frequencies in the range $[\omega_0, \omega_0 + d\omega_0]$. Then, the total dipole moment per unit volume at frequency ω is given by

$$P(\omega, r) = [\int N(\omega_0, r)e^2[m(\omega_0, r)(\omega_0^2 - \omega^2) + i\gamma(\omega_0, r)]^{-1}d\omega_0]E_0(\omega, r)$$

which results in a permittivity tensor given by

$$\epsilon(\omega, r) = \epsilon_0 I_3 + \int N(\omega_0, r)e^2[m(\omega_0, r)(\omega_0^2 - \omega^2) - i\gamma(\omega_0, r)]^{-1}d\omega_0$$

In this formula, $m(\omega_0, r)$ and $\gamma(\omega_0, r)$ denote respectively the mass tensor and the damping coefficient tensor for an electron whose nucleus is located at r and which is bound to the nucleus with a spring constant of value $K(\omega_0, r) = m(\omega_0, r)\omega_0^2$.

5.23 Temperature and field dependence of refractive index

To get temperature dependence and also field dependence of the refractive index, we have to solve Schrodinger's equation for the electron bound to the nucleus with a Gibbs distribution over the different energy eigenstates.

Specifically suppose that we solve Schrodinger's equation for the unperturbed atom having a single electron. Let its stationary energy eigenstates be denoted by $|u_n(r)>, n = 1, 2, ...$ Let E_n denote the energy of the eigenstate $u_n(r)$. Then, the initial mixed state of the system at temperature T is given by

$$\rho_0(T) = \sum_{n \geq 1} p_n(T)|u_n >< u_n|$$

or more precisely, in the position representation, it is given by

$$\rho_0(T)(r, mr') = \sum_{n \geq 1} p_n(T)u_n(r)u_n(r')^*$$

where

$$p_n(T) = exp(-\beta E_n)/Z(\beta), \beta = 1/kT, Z(\beta) = \sum_n exp(-\beta E_n)$$

Next, we solve the quantum Boltzmann's kinetic transport equation with this initial condition in the presence of an external electromagnetic field (E, B) to obtain the one particle mixed state at time τ as a function of this external field :

$$\rho_\tau(T) = \mathcal{N}(\tau, E, B\rho_0(T))$$

where $N(.)$ is a nonlinear operator obtained by solving the quantum Boltzmann equation and this nonlinear operator is applied to the initial state $\rho_0(T)$. So after time τ, we can evaluate the average electric and magnetic dipole moment of the electron in this state as a function of the temperature and the electromagnetic field and on noting that the strength of the electomagnetic field is a function of the wavelength/frequency, we obtain the quantum averaged polarization and magnetization as a function of temperature and wavelength. The cumulative distribution function of the refractive index can also be derived using our mixed state at time τ. In fact, if X is any observable, its cumulative distribution function in the state ρ is given by

$$F_\rho(x) = Tr[\rho.\chi_{(-\infty,x]}(X)]$$

where $\chi_E(x)$ is the indicator function of the set $E \subset \mathbb{R}$.

Relating the refractive index of a material to the metric tensor of space-time. The curvature of space-time affects quantum phenomena. For example, in order to take into account the space-time curvature, we have to write down Dirac's equation in curved space-time and then formulate the quantum Boltzmann equation by starting from such a generalized Dirac equation. This is accomplished as follows. Let V_a^μ be a tetrad basis for our curved space-time and let $\Gamma_\mu = \Gamma^\mu_{ab}[\gamma^a, \gamma^b]$ denote the spinor connection of the gravitational field. Then, the four component wave function satisfies

$$[V_a^\mu \gamma^a (i\partial_\mu + eA_\mu + i\Gamma_\mu) - m]\psi(x) = 0$$

From this equation, we can infer what the generalized Dirac Hamiltonian must be. This is achieved by separating the time derivative component from the spatial derivative components:

$$iV_a^0\gamma^a\partial_0\psi + [iV_a^r\gamma^a\partial_r + eV_a^\mu\gamma^a A_\mu + eV_a^\mu\gamma^a A_\mu + iV_a^\mu\gamma^a\Gamma_\mu - V_a^\mu\gamma^a m]\psi = 0$$

Now multiplying both sides of this equation by $V_b^0\gamma^b$ and using

$$V_a^0 V_b^a \gamma^a \gamma^b = (1/2)\eta^{ab} V_a^0 V_b^0 = (1/2)g^{00}$$

where η^{ab} is the Minkowski metric of flat space-time and $g^{\mu\nu}$ is the exact contravariant metric of our curved space-time, we get

$$ig^{00}\partial_0\psi + [iV_b^0 V_a^r\gamma^b\gamma^a\partial_r + eV_b^0 V_a^\mu\gamma^b\gamma^a A_\mu + iV_b^0 V_a^\mu\gamma^b\gamma^a\Gamma_\mu - V_b^0 V_a^\mu\gamma^b\gamma^a m]\psi = 0$$

This equation can be expressed in the standard Hamiltonian form by defining the curved space time Dirac Hamiltonian in an electromagnetic field as

$$H = (g^{00})^{-1}V_b^0 V_a^r\gamma^b\gamma^a(-i\partial_r) - (g^{00})^{-1}(eV_b^0 V_a^\mu\gamma^b\gamma^a A_\mu) + (g^{00})^{-1}V_b^0 V_a^\mu\gamma^b\gamma^a(-i\Gamma_\mu)$$

$$+(g^{00})^{-1}V_b^0 V_a^\mu\gamma^b\gamma^a m]$$

As an example of this calculation, consider the Schwarzchild metric in which

$$g_{00} = \alpha(r) = 1 - 2m/r, g_1 = -\alpha(r)^{-1}, g_{22} = -r^2, g_{33} = -r^2 sin^2(\theta)$$

We have

$$d\tau^2 = g_{\mu\nu}dx^\mu dx^\nu = (\omega_0)^2 - \omega_1^2 - \omega_2^2 - \omega_3^2$$

where

$$\omega_0 = \sqrt{\alpha(r)}dt, \omega_1 = \sqrt{\alpha(r)^{-1}}dr,$$
$$\omega_2 = rd\theta, \omega_3 = rsin(\theta)d\phi$$

Thus, since

$$g^{\mu\nu} = \eta^{ab}V_a^\mu V_b^\nu, g_{\mu\nu} = \eta_{ab}V_\mu^a V_\nu^b$$

we get

$$d\tau^2 = \eta_{ab}V_\mu^a V_\nu^b dx^\mu dx^\nu$$
$$= (V_\mu^0 dx^\mu)^2 - (V_\mu^1 dx^\mu)^2 - (V_\mu^2 dx^\mu)^2 - (V_\iota^3 dx^\mu)^2$$

Thus,

$$\omega_0 = \sqrt{\alpha(r)}dt = V_\mu^0 dx^\mu = V_0^0 dt + V_1^0 dr + V_2^0 d\theta + V_3^0 d\phi,$$

so that

$$\sqrt{\alpha(r)} = V_0^0, V_1^0 = V_2^0 = V_3^0 = 0,$$

5.24 Quantum statistical field theory

Let $\psi(t,r)$ denote the wave operator field of second quantized matter. The second quantized Hamiltonian in the Dirac picture is given by

$$H = \int \psi(t,r)^*((\alpha, -i\nabla + eA(r)) + \beta m)\psi(t,r)d^3r + \int V_{\mu\nu}(r,r')\psi(t,r)^*\alpha^\mu\psi(t,r)\psi(t,r')^*\alpha^\nu\psi(t,r')$$

just as in the Hartree-Fock theory. The wave operator fields satisfy the canonical equal time anticommutation relations

$$\{\psi_l(t,r), \psi_m(t,r')^*\} = \delta_{lm}\delta^3(r - r')$$

and the second term in the second quantized Hamiltonian represents the interaction between Dirac charges and currents at two different spatial points. The wave operator fields $\psi(t,r)$ evolve according to the Heisenberg dynamics

$$\partial_t\psi(t,r) = i[H, \psi(t,r)]$$

The electronic polarization operator field is given by

$$P(t,r) = -e\mathbf{r}\psi(t,r)^*\psi(t,r)$$

This is the dipole moment operator per unit volume. Let ρ_0 denote the initial state of the quantum system, say the Gibbs state:

$$\rho_0 = exp(-\beta H)/Z(\beta), Z(\beta) = Tr(exp(-\beta H)), \beta = 1/kT$$

Since we are adopting the Heisenberg picture, this state does not evolve with time, only the observables evolve with time. The average Polarization of the medium at time t is therefore given by

$$< P > (t,r) = Tr(\rho_0.P(t,r))$$

The magnetic dipole moment operator field per unit volume is given by

$$M(t,r) = \psi(t,r)^*(-e(\mathbf{L} + g\sigma)/2m)\psi(t,r)$$

and its average value is given by

$$< M > (t,r) = Tr(\rho_0 M(t,r))$$

To calculate these averages, we must first determine the dynamics of the temperature Green's function

$$G(t,r|t',r') = Tr(\rho_0 T\{\psi(t,r)\psi(t',r')^*\})$$

where T is the time ordering operator. Note that since the magnetic vector potential is not assumed to vary with time, the total Hamiltonian operator is a constant of the motion and hence so is the density operator ρ_0. Note that from the canonical anticommutation rules,

$$[\psi(t,r')^*\alpha^\mu\psi(t,r'), \psi_k(t,r)] =$$

$$= [\alpha^\mu(l,m)\psi_l(t,r')^*\psi_m(t,r'), \psi_k(t,r)] =$$

$$-\alpha^\mu(l,m)\delta_{lk}\delta^3(r'-r)\psi_m(t,r')$$

$$= -\delta^3(r-r')\alpha^\mu(k,m)\psi_m(t,r) = -\delta^3(r-r')[\alpha^\mu\psi(t,r)]_k$$

Equivalently, in vector notation,

$$[\psi(t,r')^*\alpha^\mu\psi(t,r'), \psi(t,r)] = -\delta^3(r-r')\alpha^\mu\psi(t,r)$$

It follows that if we define

$$J^\mu(t,r) = \psi(t,r)^*\alpha^\mu\psi(t,r)$$

then

$$[J^\mu(t,r'), \psi(t,r)] = -\delta^3(r-r')\alpha^\mu\psi(t,r)$$

and therefore,

$$[J^\mu(t,r')J^\nu(t,r''), \psi(t,r)] =$$

$$-[\delta^3(r-r'')J^\mu(t,r')\alpha^\nu\psi(t,r) + \delta^3(r'-r)\alpha^\mu\psi(t,r)J^\nu(t,r'')]$$

Then,

$$[\int V_{\mu\nu}(r',r'')J^\mu(r')J^\nu(r'')d^3r'd^3r'', \psi(t,r)] =$$

$$-[\int V_{\mu\nu}(r',r)J^{\mu}(t,r')d^3r']\alpha^{\nu}\psi(t.r)$$

$$-\int \alpha^{\mu}\psi(t,r)[\int V_{\mu\nu}(r,r')J^{\nu}(t,r')d^3r']$$

We may assume without loss of generality that

$$V_{\mu\nu}(r,r') = V_{\nu\mu}(r',r)$$

and then deduce that

$$[\int V_{\mu\nu}(r',r'')J^{\mu}(r')J^{\nu}(r'')d^3r'd^3r'', \psi(t,r)] =$$

$$= -\{$$

Glossary of symbols:

$A_\mu(x)$ Covariant components of the electromagnetic four potential.

$A^\mu(x)$ Contraviariant components of the electromagnetic four potential.

$F_{\mu\nu}(x)$ Covariant components of the antisymmetric electromagnetic field tensor. $F_{0r} = -F_{r0}, r = 1,2,3$ are the electric field components while F_{12}, F_{23}, F_{31} are the magnetic field components.

$\rho(t)$ density matrix representing a mixed state of a quantum system at time t.

$\rho(t,r,r') = < r|\rho(t)|r' >$ position space representation of the density matrix of a mixed state of a quantum system.

$\rho_{12..N}(t)$ Joint mixed state of N particles of a quantum system.

$Tr_{23...N}\rho_{12...N}(t) = \rho_1(t)$ marginal mixed state of the first particle of an N particle quantum system. $Tr_{23...N}$ denotes the partial trace operation. In the position space representation,

$$[Tr_{23...N}\rho_{12..N}](t,r_1,r_1') = \int \rho_{12...N}(t,r_1,r_2...,r_N,r_1',r_2...,r_N)d^3r_2...d^3r_N$$

or equivalently in terms of countable orthonormal bases,

$$< e_{i_1}|[Tr_{23...N}\rho_{12...N}](t)|e_{j_1} >=$$

$$\sum_{i_2,...,i_N} < e_{i_1} \otimes e_{i_2} \otimes ... \otimes e_{i_N}|\rho_{12...N}|e_{j_1} \otimes e_{2,i_2} \otimes ... \otimes e_{N,i_N} >$$

$p(t)$:Electric dipole moment of a single charge

$P(t,r)$ polarization or polarization operator field, ie, electric dipole moment per unit volume.

$m(t)$ magnetic dipole moment of a single charge.

$M(t,r)$ magnetization or magnetization operator field, ie magnetic moment per unit volume.

$\otimes$ tensor product between Hilbert spaces, or between vectors in two or more Hilbert spaces or between operators acting in several Hilbert spaces.

$\epsilon(\omega, r)$: permittivity tensor in the frequency domain. It relates the electric field and the polarization fields via the equation

$$P(\omega, r) = (\epsilon(\omega, r) - \epsilon_0 I)E(\omega, r)$$

$\mu(\omega, r)$:Permeability tensor in the frequency domain. It relates the magnetic field and the magnetization fields via the equation

$$\mu(\omega, r)H(\omega, r) = \mu(\omega, r)(B(\omega, r)/\mu_0 - M(\omega, r)) = B(\omega, r)$$

5.25 Root space decompositions of the complex classical Lie algebras

[a] $\mathfrak{g} = \mathfrak{sl}(n, \mathbb{C})$. This Lie algebra consists of all $n \times n$ complex matrices having trace zero. It has the root space decomposition

$$\mathfrak{g} = \mathfrak{h} \oplus \bigoplus_{k>l} \mathfrak{g}_{kl} \oplus \bigoplus_{k<l} \mathfrak{g}_{kl}$$

where $\mathfrak{h}$ consists of all complex diagonal matrices of the form

$$H = diag[h_1, h_2, ..., h_n], h_1 + h_2 + ... + h_n = -0$$

and for $k \neq l$, $\mathfrak{g}_{kl} = \mathbb{C}.E_{kl}$. Note that we can write

$$H = \sum_{k=1}^{n-1} h_k(E_{kk} - E_{nn})$$

and hence

$$\dim \mathfrak{h} = n - 1$$

and $\mathfrak{h}$ is spanned by the linearly independent elements $E_{k,k} - E_{k+1,k+1}, k = 1, 2, ..., n-1$.

We've sequentially established that if $\mathfrak{g}$ is a complex SSLA and $\mathfrak{h}$ a CSA, then $\mathfrak{h}$ is a maximal Abelian subalgebra of $\mathfrak{g}$ whose elements are all semisimple and we have the root space decomposition of the form

$$\mathfrak{g} = \mathfrak{h} \oplus \bigoplus_{\alpha \in \Delta} \mathfrak{g}_\alpha$$

where Δ is a finite subset of $\mathfrak{h}^*$ not containing the zero element and

$$\mathfrak{g}_\alpha = \{X \in \mathfrak{g} : [H, X] = \alpha(H)X \forall H \in \mathfrak{h}\}$$

and further that using the obvious identity

$$< H, H' >= Tr(ad(H).ad(H')) = \sum_{\alpha \in \Delta} dim(\mathfrak{g}_\alpha)\alpha(H)\alpha(H'), H, H' \in \mathfrak{h}$$

and the fact that (a) $<.,.>$ is non-singular on $\mathfrak{h} \times \mathfrak{h}$ and (b) that

$$<\beta, \alpha>= \beta(H_\alpha) = q_{\beta,\alpha}\alpha(H_\alpha) = q_{\beta,\alpha} < \alpha, \alpha >$$

where $q_{\beta,\alpha}$ are real rational numbers that $span_{\mathbb{C}}\Delta = \mathfrak{h}^*$ and hence that $\alpha(H_\alpha) =< \alpha, \alpha >$ is a nonzero complex number and hence also a positive rational number for any $\alpha \in \Delta$. In order to establish (b), we had defined

$$V = V_{\beta,\alpha} = \bigoplus_{k=-p}^{q} \mathfrak{g}_{\beta+k\alpha}$$

where the direct sum is over a maximal string of root spaces so that V is invariant under

$$\mathfrak{sl}(2, \mathbb{C}) = \{\bar{H}_\alpha, X_\alpha, X_{-\alpha}\}$$

and hence that

$$0 = Tr([ad(X_\alpha), ad(X_{-\alpha})]|_V) = Tr(ad(\bar{H}_\alpha)|_V)$$

and hence that

$$0 = Tr[ad(H_\alpha)|_V] = \sum_{k=-p}^{q} dim(\mathfrak{g}_{\beta+k\alpha}) < \beta + k\alpha, \alpha >$$

Fundamental to all this was our observation that $-\Delta = \Delta$. After that, we had taken a connected component C of $\mathfrak{h}'$(all the regular elements in $\mathfrak{h}$) and defined $P(C) = P$ to be the set of all roots α which assume positive values on C. (We had noted that on each connected component of $\mathfrak{h}'$, every root assumes non-zero values). Then we had defined the simple system $S(C) = S$ associated to P to be the set of all roots in P which could not be expressed as a sum of two roots in P. Based on this definition, we had noted that $\Delta = P \cup -P$ and that every root in P could be expressed as a non-negative integer linear combination of the elements of S. We had then observed that since, in particular, any element of P is a linear combination of elements of S it followed that every root (ie element of Δ) is also in particular a linear combination of elements of S and hence since $span_{\mathbb{C}}\Delta = \mathfrak{h}^*$, that $dim\mathfrak{h} \leq \mu(S)$ and hence in order to show that $\mu(S) = dim\mathfrak{h}$, it would suffice to prove the linear independence of S. This we had established by first proving that S is linearly independent over $\mathbb{R}$ which was settled using the theorem that if $v_1, ..., v_n$ is a set of vectors in a vector space V such that for any $i \neq j$, $< v_i, v_j >< 0$ then $\sum_{i=1}^n c(i)v_i = 0$ for some real numbers $\{c(i)\}$ implied $c(i) = 0$ for all i, ie, $v_1, ..., v_n$ was linearly independent over $\mathbb{R}$. Then we were in fact able to show that the $v_i's$ were in fact linearly independent over $\mathbb{C}$ by noting that if $d(i)$ were complex numbers such that $\sum_i d(i)v_i = 0$, then $\sum_i d(i) < v_i, v_j >= 0$ for all j and hence $\sum_i Re(d(i)) < v_i, v_j >= 0$ and $\sum_i Im(d(i)) < v_i, v_j >= 0 \forall j$ which implied by considering appropriate linear combinations over j using $Re(d(j))$ and $Im(d(j))$ that $\sum_i Re(d(i))v_i = \sum_i Im(d(i))v_i = 0$ and therefore using the linear independence of the $v_i's$ over

$\mathbb{R}$, we got that $d(i) = 0 \forall i$. For this result to be valid, it was noted that we did not even require $<.,.>$ to be a positive definite inner product on V. It was sufficient to require that $<.,.>$ be a positive definite bilinear form over the real vector space spanned by the $v_i's$. In order to apply this result to prove the linear independence of S, we had first shown that if $\alpha, \beta \in S$ were distinct, then $<\alpha, \beta><0$. This in turn was seen to be a consequence of the fact that if $\alpha, \beta \in S$, then $\beta - \alpha$ could not be a root and hence the maximal chain of roots $\beta + k\alpha, -p \le k \le q$ had to have $p = 0$, implying that the least eigenvalue of $ad(\bar{H}_\alpha)$ on this irreducible subspace for the Lie algebra $sl(2,\mathbb{C})$ whose standard generators are $\{\bar{H}_\alpha, X_\alpha, X_{-\alpha}\}$ is $\beta(\bar{H}_\alpha) = 2 <\beta,\alpha> / <\alpha,\alpha>$ and hence from the general theory of irreducible representations of $\mathfrak{sl}(2,\mathbb{C})$ it follows that this is a negative integer and in particular, $<\beta,\alpha>$ is a negative rational number. Further, in order to apply this result, we had first considered the real vector space $\mathfrak{h}_R = \sum_{\alpha\in\Delta} \mathbb{R}.H_\alpha$ and noted that the Cartan-Killing form defined a positive definite inner product on this real vector space. In order to show this, we had taken an $H \in \mathfrak{h}_R$ and computed

$$<H,H>= \sum_{\alpha\in\Delta} dim(\mathfrak{g}_\alpha)\alpha(H)^2$$

and noted that since $\alpha(H_\beta)$ was a real number (in fact a rational number) for any pair of roots α, β, it follows that $\alpha(H)$ was real and hence $<H,H>$ was non-negative and was zero iff $\alpha(H) = 0 \forall \alpha \in \Delta$ which implied that $H = 0$ since $span_{\mathbb{C}}\Delta = \mathfrak{h}^*$.

Some other remarks: We had stated that if for two roots $\alpha, \beta, \beta + k\alpha, -p \le k \le q$ was a maximal chain of roots, then the vector space $V = \bigoplus_{k=-p}^{q} \mathfrak{g}_{\beta+k\alpha}$ was invariant and infact even irreducible under the adjoint representation of the $\mathfrak{sl}(2,\mathbb{C})$ Lie algebra $\{\bar{H}_\alpha, X_\alpha, X_{-\alpha}\}$. Invariance is obvious and irreducibility followed from the following argument: Suppose W_1 and W_2 were two invariant irreducible subspaces of V under this representation with zero intersection. Then, from the basic properties of irreducible representations of $\mathfrak{sl}(2,\mathbb{C})$, $ad(\bar{H}_\alpha)$ has a maximal eigenvalue of j_k and a minimal eigenvalue of $-j_k$ on W_k for $k = 1,2$ where $j_k, k = 1,2$ are positive integers. We may assume without loss of generality that $j_1 \ge j_2$. Then W_1 is spanned by eigenvectors of $ad(\bar{H}_\alpha)$ with eigenvalues $j_1, j_1 - 2, ..., -j_1$ and W_2 by eigenvectors of $ad(\bar{H}_\alpha)$ with eigenvalues $j_2, j_2 - 2, .., -j_2$ and for each eigenvalue of $ad(\bar{H}_\alpha)$ in W_1, there is only one eigenvector upto a proportionality constant because of irreducibility and likewise for W_2. We had then observed that $dim\mathfrak{g}_{\beta+k\alpha} = 1, -p \le k \le q$. We next observe that the only eigenvectors of $ad(\bar{H}_\alpha)$ are precisely those in $\mathfrak{g}_{\beta+k\alpha}, k = -p, ..., q$ and the corresponding eigenvalues differ by even integers since these eigenvalues are $(\beta + k\alpha)(\bar{H}_\alpha) = \beta(\bar{H}_\alpha) + 2k, k = -p, ..., q$. For W_1 and W_2 to be disjoint apart from the zero vector, it was necessary that j_1 and j_2 differ by an odd integer for otherwise they would have a common non-zero element. But this would mean that $ad(\bar{H}_\alpha)$ has two eigenvalues that differ by an odd integer which is not possible and thereby establishing the irreducibility of V. Note that for this argument to work, we had to prove that $dim\mathfrak{g}_\alpha = 1$ forall $\alpha \in \Delta$. But this follows by considering the finite dimensional vector space

$$V_1 = span\{X_{-\alpha}) \oplus \mathfrak{h} \oplus \mathfrak{g}_\alpha \oplus \mathfrak{g}_{2\alpha} \oplus ...$$

noting that V_1 is invariant under $ad\{\bar{H}_\alpha, X_\alpha, X_{-\alpha}\}$ and hence that

$$0 = Tr([adX_\alpha, adX_{-\alpha}]|_{V_1}) = Tr(ad\bar{H}_\alpha|_{V_1})$$

$$= -alpha(\bar{H}_\alpha) + 0 + \sum_{k\geq 1} k\alpha(\bar{H}_\alpha)dim(\mathfrak{g}_{k\alpha})$$

$$= -2 + \sum_{k\geq 1} 2.dim(\mathfrak{g}_{k\alpha})$$

and therefore,

$$dim\mathfrak{g}_\alpha = 1, dim\mathfrak{g}_{k\alpha} = 0, k > 1$$

Note that this argument works only because we have been able to show that $< \alpha, \alpha >= \alpha(H_\alpha) \neq 0$.

Let C be a Weyl chamber and let P be the positive system of roots associated to this chamber and S the corresponding simple system of roots. We wish to prove that every element of P is a non-negative integer linear combination of elements from S. LetαP and let $H \in C$. Suppose $\alpha \in S$. Then there is nothing to prove. Suppose then that $\alpha \notin S$. Then $\alpha = \beta + \gamma$ for some $\beta, \gamma \in P$. Then, $\alpha(H) = \beta(H) + \gamma(H)$ shows that $\beta(H), \gamma(H)$ are positive and strictly smaller than $\alpha(H)$ (Note that all roots in P are strictly positive on C and hence strictly positive when evaluated at H). Thus, if we choose $\alpha \in P$ so that α is not expressible as a non-negative integer linear combination of elements of S and its value at H is a minimum subject to this constraint, we get a contradiction since this hypothesis implies that β, and γ are expressible as non-negative integer linear combinations of elements of S and hence so also is α. This forces us to conlcude that every element of P is expressible as a non-negative integer linear combination of elements of S.

Cartan integers: Let $S = \{\alpha_1, ..., \alpha_l\}$ be the set of simple roots relative to a positive system. In order to develop the classification theory of simple Lie algebras over $\mathbb{C}$, we have to classify only connected Dynkin diagrams because a connected Dynkin diagram or a scheme is associated only to irreducible Cartan matrices. The Cartan integers are $a(i,j) = 2 < \alpha_i, \alpha_j > / < \alpha_j, \alpha_j >, i, j = 1, 2, ..., l$. These are negative integers. We say that the Cartan matrix $A = ((a(i,j))) \in \mathbb{Z}^{l \times l}$ is irreducible iff we cannot find non-empty disjoint sets S_1, S_2 such that $S = S_1 \cup S_2$ and such that $< \alpha, \beta >= 0$ for all $\alpha \in S_1, \beta \in S_2$, ie, S_1 and S_2 are orthogonal. The main result of this section is that a Cartan matrix is irreducible iff the Lie algebra (semisimple) is simple, ie, not expressible as a direct sum of non-trivial ideals. Equivalently, a semisimple Lie algebra is expressible as a direct sum of non-trivial ideals iff the associated Cartan matrix is irreducible, ie, relative to a permutation of the simple roots, the Cartan matrix appears as a direct sum of at least two non-trivial blocks, ie, it is block diagonal.

To see this first assume that $\mathfrak{g}$ is not simple. Then, we can write

$$\mathfrak{g} = \mathfrak{g}_1 \oplus \mathfrak{g}_2$$

where $\mathfrak{g}_k, k = 1, 2$ are both non-zero ideals in $\mathfrak{g}$, ie

$$[\mathfrak{g}, \mathfrak{g}_k] \subset \mathfrak{g}_k, k = 1, 2$$

In particular,

$$[\mathfrak{g}_1, \mathfrak{g}_2] \subset \mathfrak{g}_1 \cap \mathfrak{g}_2 = 0$$

ie $\mathfrak{g}_1$ and $\mathfrak{g}_2$ mutually commute. Now let α be a root. Choose $0 \neq X \in \mathfrak{g}_\alpha$. Then

$$X = X_1 + X_2, X_k \in \mathfrak{g}_k, k = 1, 2$$

Let $H \in \mathfrak{h}$. Since $\mathfrak{g}_k, k = 1, 2$, are ideals, they are in particular, invariant under $ad(H)$ and hence from the equation

$$ad(H)(X) = \alpha(H)X$$

we get

$$(\alpha(H)X_1 - [H, X_1]) + (\alpha(H)X_2 - [H, X_2]) = 0$$

with

$$\alpha(H)X_k - [H, X_k] \in \mathfrak{g}_k, k = 1, 2$$

and therefore,

$$[H, X_k] = \alpha(H)X_k, k = 1, 2, H \in \mathfrak{h}$$

Since however $\mathfrak{g}_\alpha$ is one dimensional and $X \neq 0$, it must necessarily follow that one of the $X_k's, k = 1, 2$ vanishes. This proves exactly one of the alternatives

$$\mathfrak{g}_\alpha \subset \mathfrak{g}_k, k = 1, 2$$

occurs, ie, exactly one of the $X_k's$ vanishes, or equivalently, $X = X_1$ or $X = X_2$. This shows that for any $\alpha \in \Delta$, either $g_\alpha \subset \mathfrak{g}_1$ or $\mathfrak{g}_\alpha \subset \mathfrak{g}_2$ but not both.

The next observation we make is that relative to the Cartan-Killing form, $\mathfrak{g}_1 \perp \mathfrak{g}_2$. To see this, let $X_k \in \mathfrak{g}_k, k = 1, 2$. Then, since $\mathfrak{g}_k, k = 1, 2$ are mutually commuting ideals, it follows that for $X \in \mathfrak{g}_1$ $ad(X_1)ad(X_2)X = ad(X_1)[X_2, X] = 0$ and for $X \in \mathfrak{g}_2$, $ad(X_1)ad(X_2)X = [X_1, [X_2, X]] = 0$ and therefore $ad(X_1)ad(X_2)X = 0$ for all $X \in \mathfrak{g}$ proving thereby that

$$< X_1, X_2 >= Tr(ad(X_1).ad(X_2)) = 0$$

ie $\mathfrak{g}_1 \perp \mathfrak{g}_2$. Now suppose α is a root and $\mathfrak{g}_\alpha \subset \mathfrak{g}_1$. Then since $\mathfrak{g}_1$ is an ideal, it follows that $\bar{H}_\alpha = [X_\alpha, X_{-\alpha}] \in \mathfrak{g}_1$ and hence also $X_{-\alpha} = (1/2)[X_{-\alpha}, \bar{H}_\alpha] \in \mathfrak{g}_1$. The same is valid for $\mathfrak{g}_2$. Thus, we can write

$$S = S_1 \cup S_2, S_1 \cap S_2 = \phi,$$

$$\{H_\alpha, X_\alpha, X_{-\alpha}\} \subset \mathfrak{g}_k, \forall \alpha \in \mathfrak{g}_k, k = 1, 2$$

and since $\mathfrak{g}_1 \perp \mathfrak{g}_2$, we get that

$$< \alpha, \beta >=< H_\alpha, H_\beta >= 0, \alpha \in S_1, \beta \in S_2$$

and this proves that the Cartan matrix is decomposable.

Conversely, suppose that the Cartan matrix is decomposable. Thus,

$$S = S_1 \cup S_2, S_1 \neq \phi, S_2 \neq \phi, S_1 \cap S_2 = \phi,$$

$$< \alpha, \beta >= 0, \alpha \in S_1, \beta \in S_2$$

Let W_k be the group generated by $s_\alpha, \alpha \in S_k$ for each $k = 1, 2$. Let

$$\Delta_k = \bigcup_{\alpha \in S_k} W_1.\alpha, k = 1, 2$$

We first observe that if $\alpha \in S_1$ and $\beta \in S_2$, then since $< \alpha, \beta >= 0$, it follows that

$$s_\beta \alpha = \alpha - 2 < \beta, \alpha > \alpha / < \alpha, \alpha >= \alpha$$

and therefore,

$$s_\alpha = s_\beta s_\alpha s_\beta^{-1}$$

or equivalently,

$$s_\alpha s_\beta = s_\beta s_\alpha$$

ie, W_1 and W_2 commute. It follows that the Weyl group W, ie, the group generated by $s_\alpha, \alpha \in S$ is given by

$$W = W_1 W_2 = W_2 W_1$$

Further, we know that the set of all roots is given by

$$\Delta = \bigcup_{\alpha \in S} W.\alpha$$

and since we have seen that for any $\beta \in S_2, \alpha \in S_1$ $s_\beta \alpha = \alpha$ and $s_\alpha \beta = \beta$, it follows that

$$\Delta = (\bigcup_{\alpha \in S_1} W_1 W_2.\alpha) \cup (\bigcup_{\alpha \in S_2} W_2 W_1 \alpha)$$

$$= (\bigcup_{\alpha \in S_1} W_1 \alpha) \cup (\bigcup_{\alpha \in S_2} W_2 \alpha)$$

$$= \Delta_1 \cup \Delta_2$$

It is clear that $\Delta_1 \cap \Delta_2 = \phi$ for if not, then it will follow that a nontrivial linear combination of the elements of $S_1 \cup S_2 = S$ is zero which is false. Note that any $\alpha \in \Delta_1$ is an integer linear combination of the elements of S_1 with the integers all either non-negative or all non-positive and likewise for Δ_2. Further, if $\alpha \in \Delta_1, \beta \in \Delta_2$, then $< \alpha, \beta >= 0$ since α is in the span of S_1 and β is in the span of S_2 and by hypothesis $S_1 \perp S_2$. Further, if $\alpha \in \Delta_1$ then $-\alpha \in \Delta_1$ since otherwise $-\alpha \in \Delta_2$ will imply again that a non-trivial linear combination of elements of S is zero. (Recall that if α is a root, then so also is $-\alpha$). It is also clear that if $\alpha \in \Delta_1, \beta \in \Delta_2$, then $\alpha + \beta$ cannot be a root because if it is a root then it is either in Δ_1 or in Δ_2 and in either case, it would follow that a nontrivial linear combination of the elements of S is zero. For example if $\alpha + \beta \in \Delta_1$, then it would follow that β is in the linear span of Δ_1 which in turn is in the linear span of S_1 while β itself is in the linear span of S_2. Thus, we have shown that if $\alpha \in \Delta_1$ and $\beta \in \Delta_2$, then $[X_{\pm\alpha}, X_{\pm\beta}] = 0$ and further, $[H_\alpha, X_\beta] =< \beta, \alpha > X_\beta = 0$. In other words, the set $\{X_\alpha, X_{-\alpha}, H_\alpha\}$ commutes with $\{X_\beta, X_{-\beta}, H_\beta\}$ for each $\alpha \in \Delta_1$ and each $\beta \in \Delta_2$.

Now, define $\mathfrak{g}_1$ to be the linear span of $\{X_\alpha, X_{-\alpha}, H_\alpha, \alpha \in \Delta_1\}$ and likewise $\mathfrak{g}_2$ to be the linear span of $\{X_\alpha, X_{-\alpha}, H_\alpha, \alpha \in \mathfrak{g}_2\}$. Then from the above observations, it is clear that $\mathfrak{g}_1$ and $\mathfrak{g}_2$ are mutually commuting Lie subalgebras (because if $\alpha, \beta \in \Delta_1$, and $\alpha+\beta$ is a root, then $\alpha+\beta \in \Delta_1$ and likewise for Δ_2) and by the root space decomposition, their direct sum is $\mathfrak{g}$. Hence, $\mathfrak{g}_k, k = 1, 2$ are ideals whose direct sum is $\mathfrak{g}$ proving thereby that $\mathfrak{g}$ is not simple.

Cartan's classification of the complex simple Lie algebras

Let $\mathfrak{g}$ be a complex simple Lie algebra. Let $S = \{\alpha_1, ..., \alpha_l\}$ be a simple system of roots w.r.t a positive system P. We have already seen that

$$a(i,j) = 2 < \alpha_i, \alpha_j > / < \alpha_j, \alpha_j >, i \neq j, i, j = 1, 2, ..., l$$

are non-positive integers and these are called the Cartan integers. Draw a diagram called a Dynkin diagram which has l vertices labeled $\alpha_i, i = 1, 2, ..., l$. The weight of each vertex α_i is defined as a number w_i proportional to $< \alpha_i, \alpha_i >$. Further, by the Cauchy-Schwarz inequality, it follows that for any two integers i, j, $n(i,j) = a(i,j)a(j,i)$ is a non-negative integer assuming only the values $0, 1, 2, 3$. In the Dynkin diagram, we joint vertex i with vertex j using $n(i,j)$ links. The Dynkin diagram is clearly connected because the Cartan matrix $((a(i,j))) \in \mathbb{Z}_+^{l \times l}$ is irreducible because $\mathfrak{g}$ is assumed to be simple. Note that $a(i,i) = 2$. A Dynkin sub-diagram is a connected diagram obtained by retaining a subset of the vertices of the Dynkin diagram with the same number of links connecting any two of its vertices as in the original diagram.

Theorem 1: In a Dynkin subdiagram, D having m vertices, there cannot be more than $m - 1$ links. In particular a Dynkin diagram cannot have more than l links. Note that l is the number of simple roots which is the rank of the simple Lie algebra which is the dimension of the Cartan sub-algebra.

Proof: Let $\alpha_1, ..., \alpha_m$ be the vertices of a Dynkin subdiagram and let M denote the number of links in this subdiagram. Then since these are linearly independent, we have

$$0 < |\sum_{i=1}^{m} \alpha_i / |\alpha_i||^2 = m + 2 \sum_{1 \leq i < j \leq m} < \alpha_i, \alpha_j > / |\alpha_i||\alpha_j|$$

$$= n - \sum_{1 \le i < j \le m} \sqrt{n(i,j)} \le n - M$$

since $n(i,j) \ge 1$ if i and j are connected and zero $n(i,j) = 0$ otherwise.

Theorem 2: In a subdiagram, there cannot be any cycle, ie a closed loop.

Chapter 6

Models for the Refractive Index of Materials and Liquids

Abstract: We describe some mathematical models based on classical and quantum field theory and statistical field theory for explaining the refractive indices of materials. The first model proposes to describe the electromagnetic field interacting with the Dirac field of electrons and positrons by replacing the value of the electronic charge with a functional of the electromagnetic field. This idea is based on the fact that the nature of the singularity of the electromagnetic field completely describes the nature, ie location and value of the point charges in it. The solution to the resulting Dirac equation in a background electromanetic field will then give us the probability density function for the spatial location of the electron and by averaging the electron's electric and magnetic dipole moment operator with respect to this probability distribution, we can obtain formulas for the quantum averaged polarization and magnetization or equivalently the permittivity and permeability of the medium without having to describe the electronic field directly in terms of the electronic charge. This philosophy is in conformity with what many physicists believe today [Hans Van Leunen] that all properties of electrons should be derivable from the electromagnetic field itself. The second model describes a direct approach to computing the RI of a material based on Dirac's quantum mechanics for a system of N interacting particles in an external electromagnetic field. If we solve the Dirac equation using perturbation theory for a single particle in an electromagnetic field, we could then calculate the quantum averaged electric and magnetic dipole moment of the electron which would in turn enable us to determine the permittivity and permeability of the medium in terms of the electric and magnetic fields. However, this analysis does not show how the RI depends upon the temperature of the material. In order to obtain temperature dependence, we consider Dirac's quantum mechanics for an N particle system taking interparticle interactions into account apart from interaction of the particles with an external electromagnetic field and by partial tracing the mixed state Dirac equation over the other particles and then making some approximations we derive a quantum Boltzmann equation for the quantum density operator and if this equation is solved using perturbation theory with the intial state as the Gibbs state (which has temperature dependence), then the final equilibrium state in the presence of a static electromagnetic field and interparticle interactions will also depend upon temperature. When this final density matrix is used to compute quantum averages of the electric and magnetic dipole moment, we are able to explain the dependence of the RI on both the electromagnetic field and temperature. The wavelength dependence of the RI can be explained by assuming the background electromagnetic field to be black-body radiation which has the energy density of the electromagnetic field dependent upon both frequency/wavelength and temperature. The final model described in this manuscript takes into account cosmological and background gravitational effects on the refractive index of the material. Gravity affects quantum mechanics via the spinor connection of the gravitational field which has to be introduced into Dirac's equation in order to make it invariant under local Loretnz transformations and arbitrary diffeomorphisms of space-time. Thus, this general relativistic generalization of Dirac's equation gives us the dependence of the wave function on the background metric tensor of curved space-time. If this background metric is taken to the Schwarzchild metric, the wave function would depend upon the mass of the blackhole and the gravitational constant while if it is taken to be Robertson-Walker metric for an expanding homogeneous and isotropic universe, then the wave function will also depend on the radius of the universe and hence on Hubble's constant. Calculating the average electric and magnetic dipole moments w.r.t such a wave function would then yield the dependence of the RI on the radius of the expanding universe and on its curvature. By taking fine measurements of the RI, we would then in principle be able to measure Hubble's constant and hence the radius of the universe at the present epoch.

6.1 A.Quantum electrodynamics with the electronic charge expressed in terms of the quantum fields

A.Describing the value of a point charge and its location in space in terms of the electrostatic potential generated by it

B.Rewriting Dirac's relativistic wave equation for the electron interacting with a quantum electromagnetic field and the classical nuclear field without explicitly bringing in the electronic charge

Let the point charge be Q and let its location be r_0. The potential produced by it is

$$V(r) = Q/4\pi\epsilon|r - r_0|$$

Thus,

$$\nabla^2 V(r) = -Q\delta(r - r_0)/\epsilon$$

We can thus recover Q and r_0 from $V(.)$ using the formula

$$\int f(r)\nabla^2 V(r)d^3r = (-Q/\epsilon)f(r_0)$$

for any measurable function f having compact support. In particular, let $g(r)$ be another function. Then,

$$\frac{\int f(r)\nabla^2 V(r)d^3r}{\int g(r)\nabla^2 V(r)d^3r} = f(r_0)/g(r_0)$$

Therefore, if f, g are functions that are zero outside a compact subset K of $\mathbb{R}^3$ and are such that $f(x, y, z)/g(x, y, z) = x$, then

$$x_0 = \frac{\int f(r)\nabla^2 V(r)d^3r}{\int g(r)\nabla^2 V(r)d^3r}$$

Likewise y_0, z_0 can be recovered from $V(.)$. In this way, we can recover $r_0 = (x_0, y_{0,0})$ from $V(.)$ Then, Q is also determined using

$$Q = (-\epsilon/f(r_0))\int f(r)\nabla^2 V(r)d^3r$$

Now consider the problem of determining the point charges and their locations given their number from the electrostatic potential generated by them. Let these charges be $Q_1, ..., Q_n$ and let their locations be $r_1, ..., r_n$. If $V(r)$ is the electrostatic potential generated by them, then Poisson's equation gives

$$\epsilon\nabla^2 V(r) = \sum_{k=1}^{n} Q_k\delta(r - r_k)$$

and hence, if $f(r)$ is a bounded measurable function, we have that

$$\sum_{k=1}^{n} Q_k f(r_k) = \epsilon\int f(r)\nabla^2 V(r)d^3r$$

Now choose the functions $f_1(r) = x^m, f_2(r) = y^m, f_3(r) = z^m, m = 0, 1, 2, ...$ and derive from the above, the following system of equations

$$\sum_{k=1}^{n} Q_k x_k^m = \int x_k^m \nabla^2 V(r)d^3r,$$

$$\sum_{k=1}^{n} Q_k y_k^m = \int y_k^m \nabla^2 V(r)d^3r,$$

$$\sum_{k=1}^{n} Q_k z_k^m = \int z_k^m \nabla^2 V(r) d^3 r$$

for $m = 0, 1, 2, ...N - 1$. Thus we get a system of $3N$ equations which can be solved for $Q_k, x_k, y_k, z_k, k = 1, 2, ..., n$ or a least squares solution can be obtained provided that $N \geq 4n$. In this way a finite discrete charge distribution in space can be completely determined from the potential field. We could also do this using measurements of the electric field only using Gauss' law:

$$\epsilon div E(r) = \sum_{k=1}^{n} Q_k \delta(r - r_k)$$

Thus,

$$\epsilon \int f(r) div E(r) d^3 r = \sum_{k=1}^{n} Q_k f(r_k)$$

Another way to express the point charge distribution as a functional of the potential is to assume that the distance between the locations of any two charges in the set is greater than 2δ. Let $Q_1, ..., Q_n$ denote the point charges with locations $r_1, ..., r_n$ so that the charge density is

$$\rho(r) = \sum_{k=1}^{n} Q_k \delta(r - r_k)$$

with

$$|r_k - r_j| > \delta \forall k \neq j$$

Then let $B(\delta)$ denote the open ball in $\mathbb{R}^3$ with the origin as centre and radius δ:

$$B(\delta) = \{r \in \mathbb{R}^3 : |r| < \delta\}$$

Then given an arbitary point $r \in \mathbb{R}^3$, we have that $B(r, \delta) = r + B(\delta)$ can contain at most only one of the $r_k' s$. It follows that for any r,

$$-\epsilon \int_{B(r,\delta)} \nabla^2 V(r') d^3 r'$$

equals either zero or Q_k for some $k = 1, 2, ..., n$. It equals Q_k iff $|r - r_k| < \delta$. In other words, by moving the center ball $B(\delta)$ to different points, we get a result either equal to zero or Q_k and from the location of the centre of the ball, we can determine r_k upto an accuracy of δ. This result can also be stated as

$$lim_{r \to r_k, \delta \to 0} \int_{B(r,\delta)} (-\epsilon \nabla^2 V(r')) d^3 r'$$

$$= Q_k$$

Now consider an electron of charge $-e$ interacting with the atomic nucleus of charge Ze. Let $A_q(t,r)$ denote the free quantum electromagnetic field in space-time and let $\psi(t,r)$ denote the second quantized Dirac wave function of the electron. It satisfies the equation

$$[(\gamma^\mu(i\partial_\mu + eA_{q\mu}(t,r) + eA_{N\mu}(r)) - m]\psi(t,r) = 0$$

where

$$A_{N0}(r) = -Ze^2/|r|, A_{Nj}(r) = 0, j = 1,2,3$$

is the classical nuclear potential. According to our theory, A_q, the free quantum electromagnetic field does not have any singularity and hence if δ is sufficiently small, we have

$$\epsilon \int_{B(\delta)} \nabla^2(A_{q\mu}(t,r) + A_{N\mu}(r))d^3r = -Ze\delta_{\mu,0}$$

This equation then determines the electron charge $-e$ from the **total** electromagnetic potential. One way to write down Dirac's equation without introducing explicitly the electronic charge $-e$ is then to write it as

$$[\gamma^\mu(i\partial_\mu - Z^{-1}[\int_{B(\delta)} \nabla^2 A_0(t,r')d^3r')]A_\mu(t,r)]) - m]\psi(t,r) = 0$$

where

$$A_\mu = A_{q,\mu} + A_{N\mu}$$

is the total electromagnetic four potential comprising of the free field part described in term of photon creation and annihilation operators and the classical nuclear part.

Remark: It should be noted that in our formalism, the electron exists only because it is a part of the atom having an atomic nucleus. The existence of the electron without a corresponding nucleus is not meaningful.

Now come to the time varying case. Now the total electromagnetic field A_μ in the region $|r| > 0$, ie, in $\mathbb{R}^3 - \{0\}$ satisfies the wave equation

$$\nabla^2 A_\mu - \frac{1}{c^2}\partial_t^2 A_\mu = 0, \partial^\mu A_\mu = 0 - - - (2)$$

The question is , "Is it possible to derive all the consequences of conventional quantum field theory from equns (1) and (2) only ? To answer this question, let us assume first that the total electromagnetic field A_μ is given and we have to solve (1). Perturbatively solving it gives us to a first order approximation in the interaction term

$$\psi = \psi_0 + \psi_1,$$

$$[i\gamma^\mu\partial_\mu - m]\psi_0 = 0,$$

$$[i\gamma^\mu\partial_\mu - m]\psi_1 =$$

$$Z^{-1}[\int_{B(\delta)} \nabla^2 A_0(t,r')d^3r')]A_\mu(t,r)\gamma^\mu\psi_0$$

ψ_0 therefore represents the free Dirac field expressible as a superposition of electron and positron creation and annihilation operators. The first order perturbation ψ_1 to the free Dirac field is then

$$\psi_1 = Z^{-1}\int S(x-x')[\int_{B(\delta)} \nabla^2 A_0(t,r')d^3r')]A_\mu(x')\gamma^\mu\psi_0(x')d^4x' \quad --- (3)$$

We now use (3) to compute radiative corrections to the electron propagator in terms of the photon propagator:

$$< T(\psi(x)\psi(x')^*) > \approx < T(\psi_0(x)\psi_0(x')^*) > + < T(\psi_0(x)\psi_1(x')^*) >$$

$$+ < T(\psi_1(x)\psi_0(x')^*) >$$

where now

$$< T(\psi_0(x)\psi_0(x')^*) > = S_0(x-x')$$

is the free Dirac field electron propagator known to be given by

$$S_0(x-x') = K.\int (\gamma^\mu p_\mu - m)^{-1} exp(ip.x)d^4p$$

Now come to the time varying case for calculating charges and their velocities from the singularities in the electromagnetic field. A point charge Q moving along the trajectory $R(t), t \geq 0$ with non-relativistic velocity generates an electromagnetic field given approximately by

$$E(t,r) = \frac{Q(r-R(t))}{4\pi\epsilon|r-R(t)|^3},$$

$$B(t,r) = \frac{\mu QV(t)\times(r-R(t))}{|r-R(t)|^3}$$

Equivalently in terms of the Maxwell equations,

$$divE(t,r) = Q\delta(r-R(t))/\epsilon,$$

$$curlB(t,r) = \mu QV(t)\delta(r-R(t)) + \epsilon\partial_t E(t,r)$$

and hence we deduce that for a smooth function $f(r)$ of space coordinates,

$$\int f(r)divE(t,r)d^3r = (Q/\epsilon)f(R(t)),$$

$$\int f(r)curlB(t,r)d^3r - \epsilon\int f(r)\partial_t E(t,r)d^3r = \mu QV(t)f(R(t))$$

and by selecting f appropriately, it is clear from these two equations how to determine the charge and its trajectory including velocity at each time from the total electromagnetic field in space.

We apply this idea to quantize the electromagnetic field and Dirac field when the nucelus having charge $Q = Ze$ that binds the electron moves along a trajectory $R(t)$. The magnetic vector potential generated by such a nucleus is

$$A_N(t,r) = \frac{\mu Q V(t)}{4\pi|r - R(t)|},$$

and the electric scalar potential is given by

$$A_{N0}(t,r) = \frac{Q}{4\pi\epsilon|r - R(t)|}$$

in the non-relativistic approximation. Equivalently, in the non-relativistic approximation, we have

$$\nabla^2 A_N(t,r) = -\mu Q V(t)\delta(r - R(t)),$$

$$\nabla^2 A_{N0}(t,r) = -Q\delta(r - R(t))/\epsilon$$

so that for any test function $f(r)$, we have

$$\int f(r)\nabla^2 A_N(t,r)d^3r = -\mu Q V(t) f(R(t)),$$

$$\int f(r)\nabla^2 A_{N0}(t,r)d^3r = -Q f(R(t))/\epsilon$$

By taking the ratio of these two equations, we obtain the charge velocity vector

$$V(t) = (\mu\epsilon)^{-1}\frac{\int f(r)\nabla^2 A_N(t,r)d^3r}{\int f(r)\nabla^2 A_{N0}(t,r)d^3r}$$

The charge Q can be calculated in terms of the field A_{N0} by integrating the Laplacian applied to it over a small neighbourhood of its position $R(t)$. However, to do so, we require first to estimate $R(t)$ from the field. That can be done by taking $f(r) = |r|$ giving thereby

$$\int |r|.\nabla^2 A_{N0}(t,r)d^3r = -Q|R(t)|/\epsilon,$$

and then taking $f(r) = |r|^2$, we get

$$\int |r|^2\nabla^2 A_{N0}(t,r)d^3r = -Q|R(t)|^2/\epsilon$$

Eliminating $|R(t)|$ between these two equations gives us the charge as

$$Q = -\epsilon\frac{(\int |r|.\nabla^2 A_{N0}(t,r)d^3r)^2}{\int |r|^2\nabla^2 A_{N0}(t,r)d^3r}$$

$R(t)$ may now be calculated using

$$\int r\nabla^2 A_{N0}(t,r)d^3r = -QR(t)/\epsilon$$

and $V(t)$ using

$$\int \nabla^2 A_N(t,r)d^3r = -\mu QV(t)$$

If we assume that the quantum electromagnetic field $A_{q\mu}$ fluctuates rapidly in space, then its spatial average over any small open ball of finite radius will be negligible and hence we can to a good degree of approximation write

$$\int f(r)(\frac{\int_{B(r,\delta)} \nabla^2 A_0(t,r')d^3r'}{V(B(\delta))})d^3r = -Qf(R(t))/\epsilon,$$

$$\int f(r)(\frac{\int_{B(r,\delta)} \nabla^2 A(t,r')d^3r'}{V(B(\delta))})d^3r = -\mu QV(t)f(R(t))$$

where

$$A = A_N + A_q, A_0 = A_{N0} + A_{q0}$$

or equivalently,

$$A_\mu = A_{N\mu} + A_{q\mu}$$

These are respectively the total magnetic vector potential due to the nucleus and the quantum field and the total electrostatic field due to the same. Note that works because the nuclear potential has a singularity at the origin and at other spatial points, it varies slowly in space, while the quantum field is smooth thereby ensuring that

$$\frac{\int_{B(r,\delta)} A_{q\mu}(t,r')d^3r'}{V(B(\delta))} \approx 0$$

and since

$$\frac{\int_{B(r,\delta)} A_{N\mu}(t,r')d^3r'}{V(B(\delta))} \approx A_{N\mu}(t,r)$$

for small positive δ. Therefore,

$$\frac{\int_{B(r,\delta)} A_\mu(t,r')d^3r'}{V(B(\delta))} \approx A_{N\mu}(t,r)$$

Dirac's equation for the electron wave function is now expressible entirely in terms of the total electromagnetic field without even bringing in the electronic charge parameter. Formally, this equation is therefore expressible as

Relativistic considerations:Suppose that the nucleus is moving with relativistic velocities. Then, we replace the Laplacian operator by the wave operator in the above equations thereby obtaining

$$\square A_N(t,r) = -\mu QV(t)\delta(r - R(t)),$$

$$\Box A_{N0}(t,r) = -Q\delta(r - R(t))/\epsilon$$

where

$$\Box = \nabla^2 - \mu\epsilon\partial_t^2$$

It is then clear how all the parameters of the moving nucleus, namely it charge, position trajectory and velocity can be computed as functions of weighted integrals of the total electromagnetic field. Specifically, we find that

$$\int f(r)(\frac{\int_{B(r,\delta)} \Box A(t,r')d^3r'}{V(B(\delta))}d^3r = -\mu Q V(t)f(R(t))$$

$$\int f(r)(\frac{\int_{B(r,\delta)} \Box A_0(t,r')d^3r'}{V(B(\delta))}d^3r = -Qf(R(t))/\epsilon$$

The Dirac equation is now of the form

$$[\gamma^\mu(i\partial_\mu + F(A_{nu}(t,r), r \in \mathbb{R}^3)A_\mu(t,r)) - m]\psi(t,r) = 0$$

where $e = F(A_\nu(t,r), r \in \mathbb{R}^3)$ is the electronic charge value determined as above as spatial functional of the electromagnetic field. The electromagnetic field on the other hand satisfies Maxwell's equations in the form

$$\Box A_\mu(t,r) = 0, r \neq R(t)$$

The whole point of this exercise is that by measuring data about the Dirac wave function, or equivalently the Dirac four current density, we can in principle calculate the electromagnetic field A_μ and hence from the singularity theory mentioned above, calculate the nuclear charge as well as its trajectory.

6.2 Calculating the masses of N gravitating particles and their positions and their trajectories from measurement of the gravitational potential distribution in space-time using the Newtonian theory

Let $m_1, ..., m_N$ denote the masses of N point particles moving under their mutual gravitation along trajectories $r_1(t), ..., r_N(t)$. The Newtonian equations of motion are

$$r_j''(t) = \sum_{k=1,k\neq j}^{N} Gm_k(r_j - r_k)/|r_k - r_k|^3, j = 1,2,...,N$$

The gravitational potential generated by these masses is then

$$\Phi(r) = \sum_{j=1}^{N} Gm_j/|r - r_j|$$

and this potential satisfies Poisson's equation

$$\nabla^2\Phi(t,r)=4\pi G\sum_{j=1}^{N}m_j\delta(r-r_j(t))$$

Thus, we get for a test function $f(r)$,

$$\int f(r)\nabla^2\Phi(t,r)d^3r=4\pi G\sum_{j=1}^{N}m_jf(r_j(t))$$

By choosing test functions $f_1,...,f_N$ appropriately, we get the following linear system of equations for the masses given their positions:

$$\int_{j=1}^{N}f_k(r_j(t))m_j=\int f_k(r)\nabla^2\Phi(t,r)d^3r, k=1,2,...,N$$

Define the $N\times N$ matrix valued function of time

$$A(t)=((f_k(r_j(t))))_{1\leq k,j\leq N}$$

and let

$$B(t)=A(t)^{-1}=((b_{ij}(t)))$$

Then,

$$m_k=\sum_{j=1}^{N}b_{kj}(t)\int f_j(r)\nabla^2\Phi(t,r)d^3r$$

This formula will work even if the masses are functions of time. Now the $b_{jk}(t)'s$ are functions of the $r_j(t)'s$. So the $r_j(t)'s$ also have to be estimated from the potential distribution. Choose N vectors $\xi_1,...,\xi_N$ in $\mathbb{R}^3$. Then, we have

$$\int <\xi_k,r>\nabla^2\Phi(t,r)d^3r=4\pi G\sum_{j=1}^{N}m_j<\xi_k,r_j>, k=1,2,...,N$$

This is a system of N linear equations for the N masses and defining the matrix

$$((c_{kj}(r_j)))_{1\leq k,j\leq N}=C(r_1,...,r_N)=4\pi G((<\xi_k,r_j>))_{1\leq k,j\leq N}$$

gives us

$$m_k=\sum_{j=1}^{N}e_{kj}(r_1,...,r_N)\int <\xi_j,r>\nabla^2\Phi(t,r)d^3r, k=1,2,...,N$$

where

$$((e_{kj}))=C^{-1}$$

Thus we obtain the following N equations for $r_1, ..., r_N$

$$\sum_{j=1}^{N} b_{kj}(t) \int f_j(r)\nabla^2\Phi(t,r)d^3r$$

$$= \sum_{j=1}^{N} e_{kj}(r_1, ..., r_N) \int < \xi_j, r > \nabla^2\Phi(t,r)d^3r, k = 1, 2, ..., N$$

and by varying the vectors ξ_k in these equations, we can derive at least $3N$ equations for the N vectors $r_1, ..., r_N$ which can in principle be solved.

Now we address the same problem in Einsteinian gravity. The energy-momentum tensor for N point particles of masses $m_1, ..., m_N$ is given by

$$T^{\mu\nu}(x) = \sum_k m_k(-g(x))^{-1/2}\delta^3(x - x_k(t))(dx_k^\mu(t)/dt)(dx_k^\nu/d\tau_k)$$

where τ_k is the proper time for the kth particle. It is given by

$$d\tau^2 = g_{\mu\nu}(x_k(t))dx_k^\mu(t)dx_k^\nu(t)$$

where

$$x_k^0(t) = t$$

is the universal coordinate time. The Einstein field equations corresponding to this energy-momentum tensor are

$$G_{\mu\nu} = R_{\mu\nu} - (1/2)Rg_{\mu\nu} = -KT_{\mu\nu}, K = 8\pi G$$

We find that

$$\int T^{\mu\nu}(t,r)\sqrt{-g(t,r)}f(t,r)dtd^3r = \sum T^{\mu\nu}(x)\sqrt{-g(x)}f(x)d^4x$$

$$= \sum_k m_k \int f(x_k(t))v_k^\mu(t)v_k^\nu(t)d\tau_k(t)$$

where

$$v_k^\mu(t) = dx_k^\mu/d\tau_k$$

is the four velocity of the kth particle. From this equation, we can infer by choosing different functions $f : \mathbb{R}^4 \to \mathbb{R}$, the particle trajectories as functions of coordinate time as well as their masses.

6.3 The quantum Boltzmann equation for a plasma

In this section, we derive an approximate nonlinear evolution equation for the density operator of a single particle when the quantum plasma consists of N identical particles interacting with each other and also with an external electromagnetic field. The joint density operator of the N particles satisfies the quantum Liouville or Schrodinger-Von-Neumann equation with the Hamiltonian consisting of a sum of identical Hamiltonians each acting in a single particle Hilbert space plus the sum of identical pairwise interacting potentials of two particles with each one acting in the tensor product of two identical Hilbert spaces. By taking the partial trace of this quantum Liouville equation and making approximations (which in the classical Boltzmann kinetic transport theory are called the molecular chaos approximation), we derive an approximate quadratic nonlinear evolution equation for the single particle density operator in an external electromagnetic field. The single particle Hamiltonians can either be the single particle Schrodinger equation in an external electromagnetic field or a single particle Dirac Hamiltonian or even a single particle Dirac Hamiltonian in curved space-time interacting with an external electromagnetic field. The quadratic nonlinear terms which arise due to the pairwise interaction of particles represent quantum generalization of the so called "collsion term" that appears in the classical Boltzmann equation in kinetic transport theory and which are usually evaluated using classical scattering theory or more specifically using binary elastic collision theory of two particles. It should be noted that our method of deriving the quantum Boltzmann equation by partial tracing is the quantum analogue of the classical BBGKY theory in which one writes down the classical Liouville equation for the distribution function of N particles in phase space (ie, in the joint position-velocity space of all the N particles) and then integrates this equation over the phase space variables of all but the first particles and then makes the molecular chaos approximation in which the joint distribution of two particles is approximated by a product of the individual distributions.

Suppose that the joint density matrix of N particles is $\rho(123...N)$. It satisfies the Schrodinger equation

$$i\partial_t \rho_t(12...N) = [\sum_{a=1}^{N} H_a + \sum_{1\leq a<b\leq N} V_{ab}, \rho_t(12..N)]$$

In this equation, if we take a partial traced over $2, 3, ..., N$, we get

$$'i\partial_t \rho_{1t} = [H_1, \rho_{1t}] + (N-1)Tr_2[V_{12}, \rho_{12}]$$

and if we take the trace of the same over $3, 4, ..., N$, we get

$$i\partial_t \rho_{12t} = [H_1 + H_2 + V_{12}, \rho_{12t}] + (N-2)Tr_3[V_{13} + V_{23}, \rho_{123t}]$$

We write

$$\rho_{123} = (1/3)(\rho_{12} \otimes \rho_3 + \rho_{13} \otimes \rho_2 + \rho_1 \otimes \rho_{23}) + g_{123}$$

where g_{123} is small. Then, neglecting second order of smallness terms like V multiplied with g_{123} gives us the approximate equation

$$i\partial_t\rho_{12t} = [H_1+H_2+V_{12},\rho_{12}]+((N-2)/3)Tr_3[V_{13}+V_{23},\rho_{12}\otimes\rho_3+\rho_{13}\otimes\rho_2+\rho_1\otimes\rho_{23}]$$

This is a bit hard to handle. So we content ourselves with the approximation

$$\rho_{12} = \rho_1 \otimes \rho_1 + g_{12}$$

where g_{12} is small. We then get approximately,

$$i\partial_t\rho_{1t} = [H_1,\rho_{1t}] + (N-1)Tr_2[V_{12},\rho_1 \otimes \rho_1]$$

Writing

$$V_{12} = \sum_a W_{1a} \otimes W_{2a}$$

gives us

$$Tr_2[V_{12},\rho_1 \otimes \rho_2] = \sum_a Tr(\rho_1 W_{2a})[W_{1a},\rho_1]$$

and the our Boltzmann equation becomes

$$i\partial_t\rho_1 = [H_1,\rho_1] + (N-1)\sum_a Tr(\rho_1 W_{2a})[W_{1a},\rho_1]$$

Suppose we make the approximation

$$\rho_{123} = \rho_1 \otimes \rho_1 \otimes \rho_1 + g_{123}$$

where g_{123} is small. Then we get

$$i\partial_t\rho_{12t} = [H_1 + H_2 + V_{12},\rho_{12}] + (N-2)Tr_3[V_{13} + V_{23},\rho_1 \otimes \rho_1 \otimes \rho_1]$$

Even this equation is hard to manipulate further without assuming some specific form of the interaction potential V_{12}. We consider

$$\rho_{12} = \rho_1 \otimes \rho_1 + g_{12},$$

$$\rho_{123} = (1/3)(\rho_{12} \otimes \rho_3 + \rho_{13} \otimes \rho_2 + \rho_1 \otimes \rho_{23}) + g_{123}$$
$$= \rho_1 \otimes \rho_1 \otimes \rho_1 + (1/3(g_{12} \otimes \rho_3 + g_{13} \otimes \rho_2 + \rho_1 \otimes g_{23}) + g_{123}$$

We first derive a differential equation for g_{12} after neglecting second order of smallness terms:

$$i\partial_t\rho_{12} = i\partial_t\rho_1 \otimes \rho_1 + i\rho_1 \otimes \partial_t\rho_1$$
$$+i\partial_t g_{12}$$
$$= [H_1,\rho_1] \otimes \rho_1 + (N-1)Tr_2[V_{12},\rho_1 \otimes \rho_1] \otimes \rho_1 + \rho_1 \otimes [H_1,\rho_1]$$
$$+(N-1)\rho_1 \otimes Tr_2[V_{12},\rho_1 \otimes \rho_1] + i\partial_t g_{12}$$
$$= [H_1 + H_2 + V_{12},\rho_{12}] + (N-2)Tr_3[V_{12} + V_{13},\rho_{123}]$$

$$= [H_1 + H_2, \rho_1 \otimes \rho_1] + [V_{12}, \rho_1 \otimes \rho_1]$$
$$+(N-2)Tr_3[V_{13} + V_{23}, \rho_1 \otimes \rho_1 \otimes \rho_1]$$

After making the appropriate cancellations, we get

$$i\partial_t g_{12} =$$
$$= [V_{12}, \rho_1 \otimes \rho_1] + (N-2)Tr_3[V_{13} + V_{23}, \rho_1 \otimes \rho_1 \otimes \rho_1]$$
$$-(N-1)Tr_2[V_{12}, \rho_1 \otimes \rho_1] \otimes \rho_1$$
$$-(N-1)\rho_1 \times Tr_2[V_{12}, \rho_1 \otimes \rho_1]$$

Note that on writing

$$V_{12} = \sum_a W_{1a} \otimes W_{2a}$$

and using the fact that the $V_{jk}'s$ are identical copies of each other acting on different copies of the tensor product of two identical copies a Hilbert space just as the $H_k's$ are identical copies of each other acting on different copies of the same Hilbert space, we get

$$Tr_3[V_{13} + V_{23}, \rho_1 \otimes \rho_1 \otimes \rho_1]$$
$$= \sum_a [Tr(\rho_1 W_{2a})([W_{1a}, \rho_1] \otimes \rho_1 + \rho_1 \otimes [W_{1a}, \rho_1])]$$

A better approximation to the quantum Boltzmann equation can then be obtained by solving this equation for $g_{12}(t)$ and substituting it into the equation

$$i\partial_t \rho_1 = [H_1, \rho_1] + (N-1)Tr_2[V_{12}, \rho_{12}]$$
$$= [H_1, \rho_1] + (N-1)Tr_2[V_{12}, \rho_1 \otimes \rho_1 + g_{12}]$$

Formally this equation has the form

$$i\partial_t \rho_1(t) = [H_1, \rho_1(t)] + \delta.F_1(\rho_1(s), s \le t)$$

where F is an operator valued nonlinear functional of $\rho_1(s), s \le t$. This equation can be solved upto $O(\delta)$ using first order perturbation theory:

$$\rho_1(t) = U(t)\rho_1(0)U(t)^* + \delta.\int_0^t U(t-\tau)F_\tau(\rho_1(s), s \le \tau)U(t-\tau)^* d\tau$$

where

$$U(t) = exp(-itH_1)$$

If we consider the Hamiltonian to comprise of an interaction between the particles and an electromagnetic field, then we can write

$$H_1 = (p_1 + eA(t,r))^2/2m - e\Phi(t,r) \approx p_1^2/2m - e\Phi(t,r) + (e/2m)((p_1, A) + (A, p_1))$$

and the particle interaction potential as

$$V_{12} = V(|r_1 - r_2|)$$

In that case, in the position representation, we have

$$[p_1^2, \rho_1] = [p_1, \rho_1].p_1 + p_1.[p_1, \rho_1]$$

and noting that p_1 is represented by the kernel

$$p_1(r, r') = -i\nabla_r \delta^3(r - r') = i\nabla'_r \delta^3(r - r')$$

we get

$$[p_1, \rho_1](r, r') = -i\nabla_r \rho_1(r, r') + i\nabla'_r \rho_1(r, r')$$

$$[p_1, \rho_1].p_1(r, r') = \nabla_r.\nabla'_r \rho_1(r, r') - \nabla_r'^2 \rho_1(r, r')$$

and likewise,

$$p_1.[p_1, \rho_1](r, r') = \nabla_r.\nabla'_r \rho_1(r, r') - \nabla_r^2 \rho_1(r, r')$$

$$[(A, p_1), \rho_1](r, r') = [A_k p_{1k}, \rho_1](r, r') =$$

$$A_k(t, r) p_{1k} \rho_1(r, r') - \int \rho_1(r, r'') A_k(t, r'') p_{1k}(r'', r') dr''$$

$$= -i(A(t, r), \nabla_r)\rho_1(r, r') - i(\nabla'_r, A(t, r'))\rho_1(r, r'))$$

Therefore with neglect of nonlinear terms in the electromagnetic field, our position space representation of the quantum Boltzmann dynamics of the single particle density operator is given by

$$i\partial_t \rho_1(t, r, r') = (2m)^{-1}(2\nabla_r.\nabla'_r \rho_1(t, r, r') - (\nabla_r^2 + \nabla_r'^2)\rho_1(t, r, r'))$$

$$-(i/m)(A(t, r), \nabla_r)\rho_1(t, r, r') - i(\nabla'_r, A(t, r'))\rho_1(t, r, r')$$

$$-e\Phi(t, r)\rho_1(t, r, r') + e\Phi(t, r')\rho_1(t, r, r')$$

$$+ nonlinearterms.$$

Remark on the nonlinear terms:

$$Tr_2[V_{12}\rho_1 \otimes \rho_1](r_1, r'_1) =$$

$$V(r_1, r_2)\delta(r_1 - r''_1)\delta(r_2 - r'_2)\rho_1(t, r''_1, r'_1)\rho_1(t, r'_2, r_2)d^3r'_2 d^3r_2 d$$

$$= \int V(r_1, r_2)\rho_1(t, r_1, r'_1)\rho_1(t, r_2, r_2)d^3r_2$$

and likewise,

$$Tr_2[(\rho_1 \otimes \rho_1)V_{12}](r_1, r'_1) =$$

$$\int \rho_1(t, r_1, r_1')\rho_1(t, r_2, r_2)V_{12}(r_1', r_2)d^3r_2$$

In principle, using perturbation theory, this quantum Boltzmann equation can be solved for to obtain the single particle density operator kernel $\rho_1(t, r, r')$ as a function of the electromagnetic field and then the average dipole moment of the electron in this electromagnetic field will be given by

$$\mathbf{p}(t) = \int(-e\mathbf{r})\rho_1(t, \mathbf{r}, \mathbf{r})d^3r$$

and its average magnetic moment by

$$\mathbf{m}(t) = \int(-e\mathbf{L}(r, r')/2m)\rho_1(t, \mathbf{r}', \mathbf{r})d^3rd^3r'$$

where $\mathbf{L}(r, r')$ is the kernel of the orbital angular momentum operator in the position representation. This would then solve the problem of explaining the origin of permittivity and permeability of the plasma from the quantum statistical mechanical point of view.

Perturbative solution of the Boltzmann equation :

$$i\partial_t\rho_1(t) = [H_1, \rho_1(t)] + \delta(N-1)Tr_2[V_{12}, \rho_1(t)\otimes\rho_1(t)]$$

where δ is a perturbation parameter. We've also observed that an additional term can be added to this equation to improve its accuracy, namely,

$$\delta^2(N-1)Tr_2[V_{12}, g_{12}]$$

where

$$g_{12}(t) =$$

$$= -i\int_0^t [[V_{12}, \rho_1\otimes\rho_1] + (N-2)Tr_3[V_{13}+V_{23}, \rho_1\otimes\rho_1\otimes\rho_1]$$
$$-(N-1)Tr_2[V_{12}, \rho_1\otimes\rho_1]\otimes\rho_1$$
$$-(N-1)\rho_1\times Tr_2[V_{12}, \rho_1\otimes\rho_1]](s)ds$$
$$= \int_0^t F_2(\rho_1(s)\otimes\rho_1(s), \rho_1(s)\otimes\rho_1(s)\otimes\rho_1(s))ds$$

so our perturbative solution would have the form

$$\rho_1(t) = U(t)\rho_1(0)U(t)^* + \delta.\int_0^t U(t-s)F_1(\rho_1(s)\otimes\rho_1(s))U(t-s)^*ds$$

$$+\delta^2(N-1)\int_0^t U(t-s)[\int_0^s F_2(\rho_1(\tau)\otimes\rho_1(\tau), \rho_1(\tau)\otimes\rho_1(\tau)\otimes\rho_1(\tau))d\tau]U(t-s)^*ds$$

$$= \rho_1(t) = U(t)\rho_1(0)U(t)^* + \delta.\int_0^t U(t-s)F_1(\rho_1(s)\otimes\rho_1(s))U(t-s)^*ds$$

$$+\delta^2(N-1)\int_{0<\tau<s<t}[U(t-s)F_2(\rho_1(\tau)\otimes\rho_1(\tau),\rho_1(\tau)\otimes\rho_1(\tau)\otimes\rho_1(\tau))d\tau]U(t-s)^*]dsd\tau$$

where

$$F_1(\rho_1\otimes\rho_1)=(N-1)Tr_2[V_{12},\rho_1\otimes\rho_1]$$

The quantum Boltzmann equation derived from the Dirac relativistic wave equation

$$H(t)=(\alpha,-i\nabla+eA)+\beta m-e\Phi=H_0+H_I(t)$$

where

$$H_0=(\alpha,-i\nabla)+\beta m, H_I(t)=e(\alpha,A)-e\Phi$$

Let $V_{12}=V(r_1,r_2)$ be the interaction potential between two particles. Our aim is to formulate the Quantum Boltzmann Equation

$$i\partial_t\rho(t)=[H(t),\rho(t)]+(N-1)Tr_2[V_{12},\rho(t)\otimes\rho(t)]$$

in the position space representation. We first note that $\rho(t,r,r')$ for each t,r,r' is a 4×4 matrix and

$$[H_0,\rho]=[\alpha_k p_k+\beta m,\rho](r,r')$$

$$=\alpha_k(-i\partial_k\rho(t,r,r'))-i\partial'_k\rho(t,r,r')\alpha_k$$

and

$$[\beta,\rho](r,r')=\beta.\rho(t,r,r')-\rho(t,r,r')\beta$$

Further,

$$((\alpha,A)\rho)(r,r')=A_k(t,r)\alpha_k\rho(t,r,r')=(\alpha,A(t,r))\rho(t,r,r')$$

$$(\rho(\alpha,A))(r,r')=\rho(t,r,r')(\alpha,A(t,r'))$$

Thus,

$$[(\alpha,A),\rho](r,r')=(\alpha,A(t,r))\rho(t,r,r')-\rho(t,r,r')(\alpha,A(t,r'))$$

and likewise,

$$[\Phi,\rho](r,r')=(\Phi(t,r)-\Phi(t,r'))\rho(t,r,r')$$

6.4 Quantum electrodynamics in a background medium described by a permittivity and permeability function

We have to quantize the Maxwell equations

$$[\epsilon(\mu\nu\alpha\beta,x)*F^{\alpha\beta}(x)]_{,\nu}=0$$

where $*$ denotes space-time convolution and

$$F_{\mu\nu}(x)=A_{\nu,\mu}(x)-A_{\mu,\nu}(x)$$

The Lagrangian for density for the electromagnetic field is

$$L=(-1/4)F_{\mu\nu}(x).(\epsilon(\mu\nu\rho\sigma,x)*F^{\rho\sigma}(x))$$

The above Maxwell equations can be expressed in the space-time frequency domain as

$$\hat{\epsilon}(\mu\nu\alpha\beta,k)k_{\nu}\hat{F}^{\alpha\beta}(k)=0$$

where

$$\hat{F}^{\mu\nu}(k)=\int F^{\mu\nu}(x)exp(-ik.x)d^4x$$

$$\hat{\epsilon}(\mu\nu\alpha\beta,k)=\int\epsilon(\mu\nu\alpha\beta,x)exp(-ik.x)d^4x$$

where

$$k.x=k^0x^0-\sum_{r=1}^{3}k^rx^r,x^0=t,(x^r)_{r=1}^3=\mathbf{r}$$

Further,

$$\hat{F}_{\mu\nu}(k)=k_{\mu}\hat{A}_{\nu}(k)-k_{\nu}\hat{A}_{\mu}(k)$$

and so our Maxwell equations assume the form

$$\hat{\epsilon}(\mu\nu\alpha\beta,k)k_{\nu}(k^{\alpha}\hat{A}^{\beta}(k)-k^{\beta}\hat{A}^{\alpha}(k))=0$$

We can adopt the gauge condition

$$\hat{\epsilon}(\mu\nu\alpha\beta,k)k_{\nu}\hat{A}^{\beta}(k)=0$$

and then we get the generalized wave equation in the four wave vector domain

$$\hat{\epsilon}(\mu\nu\alpha\beta,k)k_{\nu}k^{\beta}\hat{A}^{\alpha}(k)=0---(1)$$

from which the dispersion relation can easily be obtained

$$det((\hat{\epsilon}(\mu\nu\alpha\beta,k)k_{\nu}k^{\beta}))_{0\le\mu,\alpha\le 3}=0---(2)$$

Note that this equation is an eighth degree polynomial equation in the k^μ and hence it will generally have eight solutions for k^0 in terms of $(k^r)_{r=1}^3$. In the special case when the medium is the vacuum, we have

$$\hat{\epsilon}(\mu\nu\alpha\beta, k) = \delta_{\mu\alpha}.\delta_{\nu\beta}$$

and the dispersion relation reduces to the standard one

$$k_\nu k^\nu = 0$$

Assume that the solution to the above dispersion relation (1) can be expressed as

$$k^0 = \omega_m(K), K = (k^r)_{r=1}^3, m = 1, 2, ..., 8$$

Also assume that a set of linearly independent eigenvectors $\hat{A}^\mu(k)$ that satisfy both the gauge condition and the dispersion relation are given by

$$e^\mu(K, s), s = 1, 2, ..., 8$$

Then it is clear that the electromagnetic four potential can be expanded as

$$A^\mu(x) = \int [a(K, s)e^\mu(K, s)exp(-ik^{(s)}.x)]d^3K$$

where the sum is over $s = 1, 2, ..., 8$ and

$$k^{(s)} = (\omega_s(K), K), K = (k^r)_{r=1}^3$$

More precisely, we can write

$$A^\mu(x) = \int [a(K, s)e^\mu(K, s)exp(-i(\omega_s(K)t - K.r))d^3K$$

In order that this be a real field we must assume that for each s,

$$a(K, s)^* e^\mu(K, s)^* = a(-K, s')e^\mu(-K, s')$$

for some s' and that for this s',

$$-\omega_s(K) = \omega_{s'}(-K)$$

The energy of this electromagnetic field can be obtained in the frequency domain as follows. First, the Lagrangian density in the four wave vector domain is

$$\hat{L} = (-1/4)\epsilon(\mu\nu\alpha\beta, k)\hat{F}_{\mu\nu}(k)\hat{F}^{\alpha\beta}(k)$$

Then the position fields in the four wave vector domain being $\hat{A}_\mu(k)$, it follows that the corresponding momentum fields in the four wave vector domain are

$$\hat{\pi}^\nu(k) = \partial\hat{L}/\partial(k_0\hat{A}_\nu(k))$$

$$= \epsilon(0\nu\alpha\beta, k)\hat{F}^{\alpha\beta}(k)$$

and hence the Hamiltonian density in the four vector domain is given by applying the wave vector domain Legendre transformation to the Lagrangian density:

$$\mathcal{H}(k) = \hat{\pi}^{\nu}(k)k_0\hat{A}_{\nu}(k) - \hat{L}(k)$$

$$= \epsilon(0\nu\alpha\beta, k)\hat{F}^{\alpha\beta}(k)k_0\hat{A}_{\nu}(k) + (1/4)\hat{\epsilon}(\mu\nu\alpha\beta, k)\hat{F}_{\mu\nu}(k)\hat{F}^{\alpha\beta}(k)$$

From these calculations, it is clear that the total field Hamiltonian, ie field energy can be expressed as

$$U = \int C_1(\mu\nu\alpha\beta, k)\hat{F}_{\mu\nu}(k)^*\hat{F}_{\alpha\beta}(k)d^3K$$

here it is understood thqat we substitute for k^0, its dispersion relation values in terms of $K = (k^r)_{r=1}^3$ and then sum up over all the roots. Here, $C_1(\mu\nu\alpha\beta, k)$ is a function of the permittivity-permeability tensor $\epsilon(\mu\nu\alpha\beta, k)$ (not a functional, simply an ordinary function). Alternately, substituting for $\hat{F}_{\mu\nu}(k)$ its value in term of $\hat{A}_{\mu}(k)$, we can express the field energy in the form

$$U = \int C(K, s)a(K, s)^*a(K, s)d^3K$$

where the function $C(K, s)$ is derived from $C_1(\mu\nu\alpha\beta, k)$ and the polarization vectors $e^{\mu}(K, s)$. Note that

$$\hat{F}_{\mu\nu}(k) = k_{\mu}\hat{A}_{\nu}(k) - k_{\nu}\hat{A}_{\mu}(k)$$

which in turn equals

$$a(K, s)(k_{\mu}e_{\nu}(K, s) - k_{\nu}e_{\mu}(K, s))\delta(k^0 - \omega_s(K))$$

the sum over s being understood.

6.5 Models for the refractive index of a material based on classical and quantum physics

[a] Classical physics:

Consider an electron of effective mass $m(r)$ and charge $e(r)$ moving around its equilibrium position w.r.t the atomic nucleus. Let $\gamma(r)$ denote the damping coefficient during the electron's motion and let $E(t, r)$ denote the applied external electric field. If $\xi(t)$ is the displacement of the electron relative to the nucleus, then we have from classical Newtonian mechanics,

$$m(r + \xi(t))\xi''(t) + \gamma(r + \xi(t))\xi'(t) + K(r + \xi(t))\xi(t) = -eE(t, r + \xi(t))/m$$

and since $\xi(t)$ is very small, we can assume that

$$K(r+\xi) \approx K(r), \gamma(r+\xi) \approx \gamma(r), m(r+\xi) \approx m(r)$$

Then if the electric field has frequency ω, we can write

$$E(t,r) = Re(E_0(r)exp(i\omega t))$$

and writing

$$\xi(t) = Re(\xi_0 exp(i\omega t))$$

we find on substituting into the equation of motion

$$[m(r)(\omega_0(r)^2 - \omega^2) + i\gamma(r)\omega]\xi_0 = -eE_0(r), (K(r)/m(r))^{1/2} = \omega_0(r)$$

so the dipole moment of the electron can be expressed in the form

$$p(t,r) = -e\xi(t) = Re(p_0(r)exp(i\omega t))$$

where

$$p_0(r) = e^2 E_0(r)/[m(r)(\omega_0(r)^2 - \omega^2) + i\gamma(r)\omega]$$

If there are $N(r)$ atoms per unit volume at r, then the polarization phasor (ie dipole moment per unit volume) is given by

$$P(\omega,r) = N(r)e^2 E_0(r)/[m(r)(\omega_0(r)^2 - \omega^2) + i\gamma(r)\omega]$$

from which, we deduce that the complex refractive index of the material as a function of the location is given by

$$n(\omega,r) = (1 + Ne^2\epsilon_0^{-1}/[m(r)(\omega_0(r)^2 - \omega^2) + i\gamma(r)\omega])^{1/2}$$

$$\approx 1 + N(r)e^2\epsilon_0^{-1}/2[m(r)(\omega_0(r)^2 - \omega^2) + i\gamma(r)\omega]$$

for low electron densities $N(r)$

Remark: The complex permittivity is given by

$$\epsilon(\omega,r) = \epsilon_0(1 + n(r))$$

so that if we separate out the real and imaginary parts as

$$\epsilon(\omega,r) = \epsilon_R(\omega,r) + i\epsilon_I(\omega,r)$$

then the true permittivity of the medium is given by $\epsilon_R(\omega,r)$ while the true conductivity is given by

$$\sigma(\omega,r) = -\omega\epsilon_I(\omega,r)$$

In case, the medium in which the electron moves is anisotropic, $m(r), \gamma(r), K(r)$ become 3×3 matrices and we get the result that the permittivity becomes a 3×3 matrix given by

$$\epsilon(\omega,r) = \epsilon_0(I_3 + N(r)e^2((K(r) - m(r)\omega^2) + i\gamma(r)\omega)^{-1})$$

This is independent of the temperature and also of the electric. However, it depends on the frequency and hence the wavelength of electromagnetic radiation. Further, this formula can be generalized to the situation in which there are $N(\omega_0, r)d\omega_0$ atoms per unit volume whose electrons are bound to them by spring constants having natural frequencies in the range $[\omega_0, \omega_0 + d\omega_0]$. Then, the total dipole moment per unit volume at frequency ω is given by

$$P(\omega, r) = [\int N(\omega_0, r)e^2[m(\omega_0, r)(\omega_0^2 - \omega^2) + i\gamma(\omega_0, r)]^{-1}d\omega_0]E_0(\omega, r)$$

which results in a permittivity tensor given by

$$\epsilon(\omega, r) = \epsilon_0 I_3 + \int N(\omega_0, r)e^2[m(\omega_0, r)(\omega_0^2 - \omega^2) + i\gamma(\omega_0, r)]^{-1}d\omega_0$$

In this formula, $m(\omega_0, r)$ and $\gamma(\omega_0, r)$ denote respectively the mass tensor and the damping coefficient tensor for an electron whose nucleus is located at r and which is bound to the nucleus with a spring constant of value $K(\omega_0, r) = m(\omega_0, r)\omega_0^2$.

To get temperature dependence and also field dependence of the refractive index, we have to solve Schrodinger's equation for the electron bound to the nucleus with a Gibbs distribution over the different energy eigenstates.

Specifically suppose that we solve Schrodinger's equation for the unperturbed atom having a single electron. Let its stationary energy eigenstates be denoted by $|u_n(r)>, n = 1, 2, ...$ Let E_n denote the energy of the eigenstate $u_n(r)$. Then, the initial mixed state of the system at temperature T is given by

$$\rho_0(T) = \sum_{n\geq 1} p_n(T)|u_n >< u_n|$$

or more precisely, in the position representation, it is given by

$$\rho_0(T)(r, mr') = \sum_{n\geq 1} p_n(T)u_n(r)u_n(r')^*$$

where

$$p_n(T) = exp(-\beta E_n)/Z(\beta), \beta = 1/kT, Z(\beta) = \sum_n exp(-\beta E_n)$$

Next, we solve the quantum Boltzmann's kinetic transport equation with this initial condition in the presence of an external electromagnetic field (E, B) to obtain the one particle mixed state at time τ as a function of this external field :

$$\rho_\tau(T) = \mathcal{N}(\tau, E, B\rho_0(T))$$

where $N(.)$ is a nonlinear operator obtained by solving the quantum Boltzmann equation and this nonlinear operator is applied to the initial state $\rho_0(T)$. So after time τ, we can evaluate the average electric and magnetic dipole moment of the electron in this state as a function of the temperature and the electromagnetic field and on noting that the strength of the electomagnetic field is a function of the wavelength/frequency, we obtain the quantum averaged polarization and magnetization as a function of temperature and wavelength. The cumulative distribution function of the refractive index can also be derived using our mixed state at time τ. In fact, if X is any observable, its cumulative distribution function in the state ρ is given by

$$F_\rho(x) = Tr[\rho.\chi_{(-\infty,x]}(X)]$$

where $\chi_E(x)$ is the indicator function of the set $E \subset \mathbb{R}$

6.6 Quantum statistical field theory

Let $\psi(t,r)$ denote the wave operator field of second quantized matter. The second quantized Hamiltonian in the Dirac picture is given by

$$H = \int \psi(t,r)^*((\alpha, -i\nabla + eA(r)) + \beta m)\psi(t,r)d^3r$$

$$+\int V_{\mu\nu}(r,r')\psi(t,r)^*\alpha^\mu\psi(t,r)\psi(t,r')^*\alpha^\nu\psi(t,r')$$

just as in the Hartree-Fock theory. The wave operator fields satisfy the canonical equal time anticommutation relations

$$\{\psi_l(t,r), \psi_m(t,r')^*\} = \delta_{lm}\delta^3(r-r')$$

and the second term in the second quantized Hamiltonian represents the interaction between Dirac charges and currents at two different spatial points. The wave operator fields $\psi(t,r)$ evolve according to the Heisenberg dynamics

$$\partial_t\psi(t,r) = i[H, \psi(t,r)]$$

The electronic polarization operator field is given by

$$P(t,r) = -e\mathbf{r}\psi(t,r)^*\psi(t,r)$$

This is the dipole moment operator per unit volume. Let ρ_0 denote the initial state of the quantum system, say the Gibbs state:

$$\rho_0 = exp(-\beta H)/Z(\beta), Z(\beta) = Tr(exp(-\beta H)), \beta = 1/kT$$

Since we are adopting the Heisenberg picture, this state does not evolve with time, only the observables evolve with time. The average Polarization of the medium at time t is therefore given by

$$< P > (t,r) = Tr(\rho_0.P(t,r))$$

The magnetic dipole moment operator field per unit volume is given by

$$M(t,r) = \psi(t,r)^*(-e(\mathbf{L}+g\sigma)/2m)\psi(t,r)$$

and its average value is given by

$$< M > (t,r) = Tr(\rho_0 M(t,r))$$

To calculate these averages, we must first determine the dynamics of the temperature Green's function

$$G(t,r|t',r') = Tr(\rho_0 T\{\psi(t,r)\psi(t',r')^*\})$$

where T is the time ordering operator. Note that since the magnetic vector potential is not assumed to vary with time, the total Hamiltonian operator is a constant of the motion and hence so is the density operator ρ_0. Note that from the canonical anticommutation rules,

$$[\psi(t,r')^*\alpha^\mu\psi(t,r'),\psi_k(t,r)] =$$

$$= [\alpha^\mu(l,m)\psi_l(t,r')^*\psi_m(t,r'),\psi_k(t,r)] =$$

$$-\alpha^\mu(l,m)\delta_{lk}\delta^3(r'-r)\psi_m(t,r')$$

$$= -\delta^3(r-r')\alpha^\mu(k,m)\psi_m(t,r) = -\delta^3(r-r')[\alpha^\mu\psi(t,r)]_k$$

Equivalently, in vector notation,

$$[\psi(t,r')^*\alpha^\mu\psi(t,r'),\psi(t,r)] = -\delta^3(r-r')\alpha^\mu\psi(t,r)$$

It follows that if we define

$$J^\mu(t,r) = \psi(t,r)^*\alpha^\mu\psi(t,r)$$

then

$$[J^\mu(t,r'),\psi(t,r)] = -\delta^3(r-r')\alpha^\mu\psi(t,r)$$

and therefore,

$$[J^\mu(t,r')J^\nu(t,r''),\psi(t,r)] =$$

$$-[\delta^3(r-r'')J^\mu(t,r')\alpha^\nu\psi(t,r) + \delta^3(r'-r)\alpha^\mu\psi(t,r)J^\nu(t,r'')]$$

Then,

$$[\int V_{\mu\nu}(r',r'')J^\mu(r')J^\nu(r'')d^3r'd^3r'',\psi(t,r)] =$$

$$-[\int V_{\mu\nu}(r',r)J^\mu(t,r')d^3r']\alpha^\nu\psi(t,r)$$

$$-\int \alpha^\mu\psi(t,r)[\int V_{\mu\nu}(r,r')J^\nu(t,r')d^3r']$$

We may assume without loss of generality that

$$V_{\mu\nu}(r,r') = V_{\nu\mu}(r',r)$$

and then deduce that

$$[\int V_{\mu\nu}(r',r'')J^\mu(r')J^\nu(r'')d^3r'd^3r'', \psi(t,r)] =$$

$$= -\{\alpha^\mu\psi(t,r), \int V_{\mu\nu}(r,r')J^\nu(t,r')d^3r'\}$$

We therefore obtain from the Heisenberg dynamics the following dynamical equation for the wave field operator $\psi(t,r)$.

$$i\partial_t\psi(t,r) = H_{D0}\psi(t,r) + \{\alpha^\mu\psi(t,r), \int V_{\mu\nu}(r,r')J^\nu(t,r')d^3r'\}$$

where

$$H_{D0} = (\alpha, -i\nabla) + \beta m$$

is the first quantized free particle Dirac Hamiltonian. We can further approximate this equation by replacing $J^\mu(t,r')$ on the rhs by its quantum average

$$< J^\mu(t,r) >= Tr(\rho_0(T)J^\mu(t,r))$$

It should be noted that this average current density can be expressed in terms of the temperature Green's function as

$$< J^\mu(t,r) >= -Tr[\alpha^\mu G(t,r|t,r')]|_{r'=r}$$

where the trace here is an ordinary matrix trace for 4×4 matrices.

Reference:Fetter and Walecka, "Quantum Theory of Many Particle Systems", Dover, 1971.

6.7 Relating the refractive index of a material to the metric tensor of space-time

The curvature of space-time affects quantum phenomena. For example, in order to take into account the space-time curvature, we have to write down Dirac's equation in curved space-time and then formulate the quantum Boltzmann equation by starting from such a generalized Dirac equation. This is accomplished as follows. Let V_a^μ be a tetrad basis for our curved space-time and let $\Gamma_\mu = \Gamma^\mu_{ab}[\gamma^a, \gamma^b]$ denote the spinor connection of the gravitational field. Then, the four component wave function satisfies

$$[V_a^\mu\gamma^a(i\partial_\mu + eA_\mu + i\Gamma_\mu) - m]\psi(x) = 0$$

From this equation, we can infer what the generalized Dirac Hamiltonian must be. This is achieved by separating the time derivative component from the spatial derivative components:

$$iV_a^0\gamma^a\partial_0\psi + [iV_a^r\gamma^a\partial_r + eV_a^\mu\gamma^a A_\mu + eV_a^\mu\gamma^a A_\mu + iV_a^\mu\gamma^a\Gamma_\mu - V_a^\mu\gamma^a m]\psi = 0$$

Now multiplying both sides of this equation by $V_b^0\gamma^b$ and using

$$V_a^0 V_b^a\gamma^a\gamma^b = (1/2)\eta^{ab}V_a^0 V_b^0 = (1/2)g^{00}$$

where η^{ab} is the Minkowski metric of flat space-time and $g^{\mu\nu}$ is the exact contravariant metric of our curved space-time, we get

$$ig^{00}\partial_0\psi + [iV_b^0 V_a^r\gamma^b\gamma^a\partial_r + eV_b^0 V_a^\mu\gamma^b\gamma^a A_\mu + iV_b^0 V_a^\mu\gamma^b\gamma^a\Gamma_\mu - V_b^0 V_a^\mu\gamma^b\gamma^a m]\psi = 0$$

This equation can be expressed in the standard Hamiltonian form by defining the curved space time Dirac Hamiltonian in an electromagnetic field as

$$H = (g^{00})^{-1}V_b^0 V_a^r\gamma^b\gamma^a(-i\partial_r) - (g^{00})^{-1}(eV_b^0 V_a^\mu\gamma^b\gamma^a A_\mu) + (g^{00})^{-1}V_b^0 V_a^\mu\gamma^b\gamma^a(-i\Gamma_\mu)$$
$$+(g^{00})^{-1}V_b^0 V_a^\mu\gamma^b\gamma^a m]$$

As an example of this calculation, consider the Schwarzchild metric in which

$$g_{00} = \alpha(r) = 1 - 2m/r, g_1 = -\alpha(r)^{-1}, g_{22} = -r^2, g_{33} = -r^2 sin^2(\theta)$$

We have

$$d\tau^2 = g_{\mu\nu}dx^\mu dx^\nu = (\omega_0)^2 - \omega_1^2 - \omega_2^2 - \omega_3^2$$

where

$$\omega_0 = \sqrt{\alpha(r)}dt, \omega_1 = \sqrt{\alpha(r)^{-1}}dr,$$
$$\omega_2 = rd\theta, \omega_3 = rsin(\theta)d\phi$$

Thus, since

$$g^{\mu\nu} = \eta^{ab}V_a^\mu V_b^\nu, g_{\mu\nu} = \eta_{ab}V_\mu^a V_\nu^b$$

we get

$$d\tau^2 = \eta_{ab}V_\mu^a V_\nu^b dx^\mu dx^\nu$$
$$= (V_\mu^0 dx^\mu)^2 - (V_\mu^1 dx^\mu)^2 - (V_\mu^2 dx^\mu)^2 - (V_\mu^3 dx^\mu)^2$$

Thus,

$$\omega_0 = \sqrt{\alpha(r)}dt = V_\mu^0 dx^\mu = V_0^0 dt + V_1^0 dr + V_2^0 d\theta + V_3^0 d\phi,$$

so that

$$\sqrt{\alpha(r)} = V_0^0, V_1^0 = V_2^0 = V_3^0 = 0,$$

Note that in these conventions,

$$(V_\mu^0) = (V_0^0, V_1^0, V_2^0, V_3^0)$$

Likewise,

$$\omega_1 = V_\mu^1 dx^\mu = \alpha(r)^{-1/2}dr$$

and therefore,

$$V_0^1 = 0, V_1^1 = \alpha(r)^{-1/2}, V_2^1 = 0, V_3^1 = 0$$

Note that

$$(V_\mu^1) = (V_0^1, V_1^1, V_2^1, V_3^1)$$

$$\omega_2 = V_\mu^2 dx^\mu = rd\theta, \omega_3 = V_\mu^3 dx^\mu = r.sin(\theta)d\phi$$

so that

$$V_0^2 = 0, V_1^2 = 0, V_2^2 = r, V_3^2 = 0,$$

$$V_0^3 = 0, V_1^3 = 0, V_2^3 = 0, V_3^3 = r.sin(\theta)$$

The spinor connection of the gravitational field is given by

$$\Gamma_\mu = V_{a\nu} V_{b:\mu}^\nu [\gamma^a, \gamma^b] = \omega_\mu^{ab} [\gamma^a, \gamma^b]$$

This can be derived in various ways, one by using the fact that the covariant derivative of the tetrad is zero, ie

$$V_{\mu,\nu}^a - \Gamma_{\mu\nu}^\rho V_\rho^a + \omega_\nu^{ab} V_{b\mu} = 0$$

or equivalently,

$$V_{\mu:\nu}^a + \omega_\nu^{ab} V_{b\mu} = 0$$

The second derivation is based on starting with the Einstein-Hilbert Lagrangian for the gravitational field, namely the curvature tensor in spinor notation:

$$R = R_{\mu\nu}^{ab} V_a^\mu V_b^\nu,$$

$$R_{\mu\nu}^{ab} = \omega_{\nu,\mu}^{ab} - \omega_{\mu,\nu}^{ab} - [\omega_\mu, \omega_\nu]^{ab}$$

and setting the variational derivative of $\int Rd^4x$ w.r.t ω_μ^{cd} to zero to arrive at an algebraic equation for ω_μ^{ab} which when solved yields the desired form of the gravitational spinor connection. The third and most interesting way to derive the form of the spinor connection of the gravitational field is based on the use of non-Abelian gauge group theory with the gauge group now being the spinor representation of the Lorentz group. In other words, the transformation law of the spinor connection of the gravitational field under a local Lorentz transformation should be such that the Dirac equation remains invariant under it. Let $\Gamma_\mu(x)$ denote the spinor connection in one frame. The corresponding covariant derivative is $\nabla_\mu = \partial_\mu + \Gamma_\mu$. Now suppose we apply a local Lorentz transformation $\Lambda(x)$. Then the Dirac wave function transforms from $\psi(x)$ to $D(\Lambda(x))\psi(x)$ and for the Dirac equation to remain invariant under this transformation, we require that $\Gamma_\mu(x)$ should transform to $\Gamma'_\mu(x)$ where

$$D(\Lambda(x))(\partial_\mu + \Gamma_\mu(x))D(\Lambda(x))^{-1} =$$

$$\partial_\mu + \Gamma'_\mu(x)$$

This is equivalent to requiring that

$$\Gamma'_{\mu}(x) = D(\Lambda(x))\Gamma_{\mu}(x)D(\Lambda(x))^{-1} + D(\Lambda(x))(\partial_{\mu}D(\Lambda(x))^{-1})$$

or equivalently that

$$\Gamma'_{\mu}(x) = D(\Lambda(x))\Gamma_{\mu}(x)D(\Lambda(x))^{-1} - (\partial_{\mu}D(\Lambda(x)))D(\Lambda(x))^{-1}$$

Equivalently, if

$$\Lambda(x) = I + \omega(x)$$

is an infinitesimal local Lorentz transformation, then we require that

$$\delta\Gamma_{\mu}(x) = \Gamma'_{\mu}(x) - \Gamma_{\mu}(x) =$$

$$[D(\omega(x)), \Gamma_{\mu}(x)] + \partial_{\mu}D(\omega(x))$$

where

$$D(\omega(x)) = (1/4)[\gamma^a, \gamma^b]\omega_{ab}(x)$$

is the differential of the spinor representation (interpreted in terms of Lie algebra representations) of the Lorentz group evaluated at $\omega(x)$. On the other hand, if we take $\Gamma_{\mu}(x) = V_{a\nu:\mu}V_b^{\nu}[\gamma^a, \gamma^b]/2$, then under this infinitesimal local Lorentz transformation, it changes by

$$\delta\Gamma_{\mu}(x) = ([\gamma^a, \gamma^b]/4)[(\delta V_{a\nu:\mu})V_b^{\nu} + V_{a\nu:\mu}\delta V_b^{\nu}]$$

where

$$\delta V_{a\nu} = \omega_a^b(x)V_{b\nu}$$

so that

$$\delta V_{a\nu:\mu} = \omega_a^b V_{b\nu:\mu} + \omega_{a,\mu}^b V_{b\nu}$$

since the ω_b^a's and ω_{ab}'s transform as scalar fields under diffeomorphisms of space-time. By comparing these two transformation laws, we can show that they are identical using the canonical commutation relations for the Loretnz algebra generators.

6.8 Cosmological effects on the refractive index

[a] Classical analysis, main idea: The metric of space time is assumed to be the Robertson-Walker metric corresponding to a homogeneous, isotropic expanding universe:

$$d\tau^2 = dt^2 - S^2(t)f(r)dr^2 - r^2(d\theta^2 + sin^2(\theta)d\phi^2)$$

so that

$$g_{00} = 1, g_{11} = -S^2(t)f(r), g_{22} = -r^2, g_{33} = -r^2 sin^2(\theta)$$

This metric describes a comoving frame, ie, a frame in which a particle at rest, ie having fixed spatial coordinates r, θ, ϕ satisfies the geodesic equations. In order to study the effect of the expanding universe on the refractive index of a body, we have to formulate the Vlasov equations for the particle distribution function and the electromagnetic field in this background metric and then compute statistical averages of the electric and magnetic dipole moments of a charge. The Boltzmann equation in this metric is expressed as

$$\partial_t f(t,\mathbf{r},\mathbf{v})+(\mathbf{v},\nabla_e)d(t,\mathbf{r},\mathbf{v})+(\mathbf{F}(t,\mathbf{r},\mathbf{v}),\nabla_v)f(t,\mathbf{r},\mathbf{v}) = (f_0(\mathbf{v})-f(t,\mathbf{r},\mathbf{v}))/\tau(\mathbf{v}$$

where

$$\mathbf{v} = d\mathbf{r}/dt, v^r = dx^r/dt, r = 1,2,3$$

and $\mathbf{F}(t,\mathbf{r},\mathbf{v})$ is determined from the equation of motion of a charged particle in the curved space-time metric of Robertson and Walker and also under the influence of an electromagentic field:

$$du^\mu/d\tau + \Gamma^\mu_{\alpha\beta}(x)u^\alpha u^\beta = eF^{\mu\nu}u_\nu$$

where

$$u^\mu = dx^\mu/d\tau, d\tau = dt(g_{00} + 2g_{0r}v^r + g_{rs}v^r v^s)^{1/2}$$

The Maxwell equations are expressed as

$$F_{\mu\nu} = A_{\nu,\mu} - A_{\mu,\nu},$$

$$F^{\mu\nu}_{:\nu} = \mu_0 J^\mu$$

where

$$J^0 = q\int f(t,\mathbf{r},\mathbf{v})d^3v, J^r = q\int v^r f(t,\mathbf{r},\mathbf{v})d^3v$$

It should be noted that if m_0 is the mass of each charge, then the energy-momentum tensor of this charged matter fluid is given by

$$T^{\mu\nu}(x) = m_0\int f(t,\mathbf{r},\mathbf{v})u^\mu u^\nu d^3v$$

where

$$u^0 = dt/d\tau = (g_{00}(t,r) + 2g_{0r}(t,r)v^r - g_{rs}(t,r)v^r v^s)^{-1/2},$$

$$u^r = dx^r/d\tau = v^r dt/d\tau = v^r u^0$$

Remarks on the quantization of the energy-momentum tensor of a system of N particles and the Einstein field equations.

[a] Let $r_1,...,r_N$ denote the position operators of N particles and let $p^\mu_k, k = 1,2,....,N$ denote their four momentum operators. Thus, $p^0_k = H_{0k}$ is the energy operator of the kth particle and $\mathbf{p}^r = (p^r_k, r = 1,2,3)$ are the Cartesian components of the three momentum vector of the kth particle. According to Dirac's relativistic theory of the electron, if the particles do not interact, then

$$H_{0k} = (\alpha,\mathbf{p}_k) + \beta m_k = H_{0k}(\mathbf{p}_k)$$

assuming that the kth particle has mass m_k. More precisely, we should use the curved space-time Dirac Hamiltonian for H_{0k} in terms of the background metric and an associated tetrad basis. We denote this Hamiltonian by $H_{0k}(\mathbf{r}_k\mathbf{p}_k, g_{\mu\nu})$.

Then the energy-momentum tensor operator field of the matter field comprising the N particles is given by

$$T^{\mu\nu} = \sum_{k=1}^{N} m_k \delta^3(r - r_k)(-g(r_k))^{-1/2} p_k^\mu p_k^\nu / H_{0k}$$

We are assuming that although the particle's motion is quantized, the metric of space-time is classical which means that the Einstein field equations are

$$G^{\mu\nu}(x) = R^{\mu\nu}(x) - (1/2)Rg^{\mu\nu}(x) = -8\pi G < T^{\mu\nu}(x) >$$

where the average $< . >$ is a quantum average that is taken with respect to the evolving wave function or mixed state of the system of N particles. In this formalism, we are working in the Schrodinger picture in which states evolve with time but observables remain constant in time. The mixed state of the system of N particles $\rho(t)$ satisfies the quantum Liouville equation:

$$i\rho'(t) = [\sum_{k=1}^{N} H_{0k}, \rho(t)]$$

where in each component $H_{0k}, k = 1, 2, ..., N$, identical copies of the same Dirac matrices are taken but acting on different tensor product Hilbert space components. Thus, the Hilbert space of the system of N particles is given by

$$\mathcal{H} = \bigotimes_{k=1}^{N} \mathcal{H}_k, \mathcal{H}_k = L^2(\mathbb{R}^3) \otimes \mathbb{C}^4$$

If all the particles are identical then we can use the Boltzmann equation approximation by considering the marginal state $\rho_1(t)$ of just one particle which may be interacting with the other particles also. $\rho_1(t)$ will approximately satisfy an equation of the form

$$i\rho_1'(t) = [H_{01}(r_1, p_1, g_{\mu\nu}), \rho_1(t)] + (N-1)Tr_2[V(r_1, p_1, r_2, p_2), \rho_1(t) \otimes \rho_1(t)]$$

With this quantum Boltzmann equation approximation, we have the following approximation for the quantum averaged energy-momentum tensor:

$$< T^{\mu\nu}(t, r) >= NTr(\rho_1(t)\delta^3(r - r_1)(-g(r)^{-1/2}[p_1^\mu p_1^\nu / H_{01}(r_1, p_1, g_{\mu\nu})]$$

Note that to keep this average real, we may interpret $p_1^\mu p_1^\nu / H_{01}$ as

$$(1/2)(p_1^\mu H_{01}^{-1} p_1^\nu + p_1^\nu H_{01}^{-1} p_1^\mu]$$

Glossary of symbols:

$A_\mu(x)$ Covariant components of the electromagnetic four potential.

$A^\mu(x)$ Contravariant components of the electromagnetic four potential.

$F_{\mu\nu}(x)$ Covariant components of the antisymmetric electromagnetic field tensor. $F_{0r} = -F_{r0}, r = 1, 2, 3$ are the electric field components while F_{12}, F_{23}, F_{31} are the magnetic field components.

$\rho(t)$ density matrix representing a mixed state of a quantum system at time t.

$\rho(t, r, r') =< r|\rho(t)|r' >$ position space representation of the density matrix of a mixed state of a quantum system.

$\rho_{12..N}(t)$ Joint mixed state of N particles of a quantum system.

$Tr_{23...N}\rho_{12...N}(t) = \rho_1(t)$ marginal mixed state of the first particle of an N particle quantum system. $Tr_{23...N}$ denotes the partial trace operation. In the position space representation,

$$[Tr_{23...N}\rho_{12..N}](t, r_1, r_1') = \int \rho_{12...N}(t, r_1, r_2..., r_N, r_1', r_2..., r_N)d^3r_2...d^3r_N$$

or equivalently in terms of countable orthonormal bases,

$$< e_{i_1}|[Tr_{23...N}\rho_{12...N}](t)|e_{j_1} >=$$

$$\sum_{i_2,...,i_N} < e_{i_1} \otimes e_{i_2} \otimes ... \otimes e_{i_N}|\rho_{12...N}|e_{j_1} \otimes e_{2,i_2} \otimes ... \otimes e_{N,i_N} >$$

$\psi(x)$: four component Dirac wave function, also called a bispinor.

$\Gamma_\mu(x)$: Spinor connection of the gravitational field. Used to determine the effect of gravity on Dirac's wave function for relativistic quantum mechanics.

$S(t)$ Radius of the expanding universe at the epoch t.

$V_a^\mu(x)$ Tetrad basis for the metric of space-time. This enables us to express the metric locally in Minkowski form. In other words, it can be used to describe a locally inertial frame.

Chapter 7

More Problems in Probability Theory, Antennas and Refractive Index of Materials

7.1 Levy's modulus of continuity for Brownian motion

Let

$$g(\delta) = \sqrt{2\delta.ln(1/\delta)}, 0 < \delta < 1$$

Let

$$0 < \theta < 1$$

Define

$$E_n = \{max_{1 \le j \le 2^n} |B(j/2^n) - B((j-1)/2^n)|/g(1/2^n) \le \sqrt{1-\theta}\}, n = 1, 2, ...$$

We shall prove that

$$\sum_n P(E_n) < \infty$$

and then it will follow from the Borel-Cantelli Lemma that

$$P(E_n, i.o) = 0$$

or equivalently that for a.e.ω, there exists a finite positive integer $N(\omega)$ such that for all $n > N(\omega)$, we have

$$max_{1 \le j \le 2^n} |B(j/2^n, \omega) - B((j-1)/2^n, \omega)|/g(1/2^n) > \sqrt{1-\theta}$$

which would imply that for such ω,

$$limsup_{h\downarrow 0}max_{0\leq t\leq 1-h}|B(t+h,\omega)-B(t,\omega)|/g(h)\geq\sqrt{1-\theta}$$

and hence letting $\theta\downarrow 0$ through rationals (Note that $P(F_n)=1, n=1,2,...$ implies $P(\bigcap_n F_n)=1$), we would get the first half of the Levy modulus of continuity theorem:

$$limsup_{h\downarrow 0}max_{0\leq t\leq 1-h}|B(t+h)-B(t)|/g(h)\geq 1 a.s$$

To prove the summability of $P(E_n)$, we note that

$$P(E_n)=(1-\xi_n)^{2^n}\leq exp(-2^n\xi_n)$$

in view of the independence of the events E_n, where

$$\xi_n=P(|B(1/2^n)|>g(1/2^n)\sqrt{1-\theta})$$
$$=1-\Phi(x_n)$$

where $\Phi(x)$ is the standard normal distribution function and

$$x_n=g(1/2^n)\sqrt{1-\theta}.2^{n/2}$$

Now, for any $x>0$, we have using integration by parts,

$$1-\Phi(x)=(2\pi)^{-1/2}\int_x^\infty exp(-u^2/2)du$$

$$=(2\pi)^{-1/2}(\int_x^\infty(1/u).u.exp(-u^2/2)du$$

$$=(2\pi)^{-1/2}(exp(-x^2/2)/x-\int_x^\infty(1/u^2)exp(-u^2/2)du)$$

$$\geq(2\pi)^{-1/2}(exp(-x^2/2)/x-x^{-2}\int_x^\infty exp(-u^2/2)du)$$

and this inequality can also be expressed as

$$(1+1/x^2)(1-\Phi(x))\geq(2\pi)^{-1/2}x^{-1}.exp(-x^2/2)$$

or equivalently,

$$1-\Phi(x)\geq\frac{x}{1+x^2}.exp(-x^2/2)$$

Thus,

$$\xi_n\geq\frac{x_n}{1+x_n^2}exp(-x_n^2/2)$$

and

$$x_n^2/2=g(1/2^n)^2(1-\theta).2^{n-1}=2.(1/2^n).log(2^n).2^{n-1}(1-\theta)$$
$$=n(1-\theta)log(2)$$

and hence,

$$x_n \to \infty$$

and further,

$$exp(-x_n^2/2) = 2^{-n(1-\theta)}$$

so that

$$exp(-2^n \xi_n) \geq K_n.exp(-2^{n\theta}), K_n = (2\pi)^{-1/2} \frac{x_n}{1+x_n^2} \leq 1$$

for large n and hence

$$\sum_n exp(-2^n \xi_n) \leq \sum_n exp(-2^{n\theta}) < \infty$$

from the desired summability of $P(E_n)$ follows.

To prove the second half of the Levy modulus theorem, we choose $\epsilon > 0$ and consider the events

$$E_n = \{max_{1 \leq i \leq i+k \leq 2^n, 0 \leq k \leq 2^{n\theta}} |B((i+k)/2^n) - B(i/2^n)|/g(k/2^n) > 1 + \epsilon\}$$

and deduce easily that

$$P(E_n) \leq 2^n \sum_{k=1}^{2^{n\theta}} P(|B(k/2^n)| > (1+\epsilon)g(k/2^n))$$

$$= 2^n \sum_{k=1}^{2^{n\theta}} P(|Z| > (1+\epsilon)g(k/2^n)(2^n/k)^{1/2})$$

where Z is a standard normal random variable. Now define

$$x(n,k) = (1+\epsilon)g(k/2^n)(2^n/k)^{1/2}$$

Then,

$$x(n,k)^2/2 = (1+\epsilon)^2(k/2^n)log(2^n/k).2^n/k = (1+\epsilon)^2 log(2^n/k)$$

$$exp(-x(n,k)^2/2) = (2^n/k)^{-(1+\epsilon)^2}$$

and hence

$$P(E_n) \leq 2^{n(1-(1+\epsilon)^2)}. \sum_{k=1}^{2^{n\theta}} k^{(1+\epsilon)^2}$$

Now,

$$\int_0^{2^{n\theta}} x^{(1+\epsilon)^2} dx = 2^{n\theta((1+\epsilon)^2+1)}/((1+\epsilon)^2+1)$$

and hence,

$$exp(-x(n,k)^2/2) \leq K.2^{n(1-(1+\epsilon)^2+n\theta(1+\epsilon)^2+n\theta}$$

$$= K.2^{n[1+\theta-(1+\epsilon)^2(1-\theta)]}$$

and hence, if we select ϵ so that

$$(1+\epsilon)^2 > (1+\theta)/(1-\theta) --- (a)$$

then it would follow that

$$\sum_n P(E_n) < \infty$$

and hence by the Borel-Cantelli lemma,

$$P(E_n, i.o) = 0$$

We then get the result that for a.e. ω, there exists a finite positive integer $N(\omega)$ such that for all $n > N(\omega)$,

$$\{max_{1\leq i\leq i+k\leq 2^n, 0\leq k\leq 2^{n\theta}}|B((i+k)/2^n) - B(i/2^n)|/g(k/2^n) \leq 1+\epsilon\}$$

and hence since $2^{n\theta}/2^n = 2^{-n(1-\theta)}$ converges to zero as $n \to \infty$, it follows from the continuity of the Brownian sample paths that

$$limsup_{h\downarrow 0} max_{0\leq t\leq 1-h}(|B(t+h) - B(t)|/g(h) \leq 1+\epsilon$$

and now letting ϵ decrease to zero yields the second half of the Levy modulus of continuity theorem.

Remark: Let $t, s \in [0,1]$, $t \geq s, t-s = h$. Let $\delta_1 > 0$ be given. Let $D_n = \{k/2^n : k = 0, 1, ..., 2^n\}$. Note that $D_n \subset D_{n+1}$. Note that $D = \bigcup_n D_n$ is the set of all dyadic rationals in $[0,1]$. Fix any $\delta > 0$ and choose n large enough so that $|t-t'| < \delta, |s-s'| < \delta$ for some $t', s' \in D_n$ where δ is chosen so that $|u| < \delta$ implies $|B(t+u) - B(t)| < \delta_1$ for all $t \in [0, 1-u]$. This is possible since B is uniformly continuous on $[0,1]$. Then

$$|B(t) - B(t')|, |B(s) - B(s')| < \delta_1$$

and hence,

$$|B(t) - B(s)| \leq 2\delta_1 + |B(t') - B(s')|$$

7.2 Test 2:Antennas and Wave Propagation

Question 3 is compulsory. Attempt any three questions from the remaining. Each question carries ten marks.

[1] Explain how using the principle of pattern multiplication, you will calculate the far field radiation pattern produced by a microstrip antenna designed as a cuboidal cavity of dimensions a, b, d with the only non-vanishing component of the magnetic vector potential being $A_z(x, y, z)$ satisfying the Helmholtz equation

$$(\nabla^2 + k^2)A_z = 0$$

with the boundary condition that the tangential components of the electric field and normal component of the magnetic field vanish on the boundaries. Derive an explicit formula for the far field radiation pattern.

[2] A planar Archimedian spiral antenna has the equation of its curve given by

$$\rho = A.exp(b\phi), z = 0$$

where (ρ, ϕ, z) are cylindrical coordinates. This antenna carries a current I at frequency ω. Calculate the radiation resistance of this antenna assuming that N complete spirals are made. It is known that if all the dimensions are generated by simply rotating an antenna, then the antenna is broadband. Justify this for the Archimedian spiral.

[3] A horn antenna consists of a rectangular waveguide with transverse dimensions a, b feeding into a horn having a spherical aperture of radius R with centre at the centre of the mouth of the waveguide assumed to be on the xy plane and extending from azimuthal angle$\theta = 0$ to $\theta = \alpha$. Determine the following:

[a] The distance $\delta(x, y, \theta)$ between a point $(x, y, 0)$ at the mouth of the guide and a point (R, θ, ϕ) on the spherical horn surface. Approximate this distance upto quadratic terms in x, y assuming that $a, b << R$.

[b] Derive expressions for the fields $E_x(x, y), E_y(x, y), H_x(x, y), H_z(x, y)$ at the waveguide mouth assuming a $TE_{m,n}$ mode.

[c] Calculate using the phase shift principle, the fields at the horn aperture using the approximations made in [a]. In other words you must note that these field are given by

$$\int_0^a \int_0^b \mathbf{E}(x, y) exp(-ik\delta(x, y, \theta)) dx dy,$$

and

$$\int_0^a \int_0^b \mathbf{H}(x, y) exp(-ik\delta(x, y\ \theta)) dx dy;$$

[d] Using the formulas for the electromagnetic fields on the horn surface derived in [c], calculate the Poynting vector pattern in the far field zone by using the fact that the surface electric current density on the horn surface is $\mathbf{J}_s = \hat{r} \times \mathbf{H}$ and the surface magnetic current density on the same is $\mathbf{M}_s = -\hat{r} \times \mathbf{E}$.

[4] Let $E(x, y)$ denote the electric field on a planar aperture D in the xy plane. justify the statement based on the equation of wave propagation that the field in space at the point (X, Y, Z) produced by this aperture must be of the form

$$F(X, Y, Z) = \int f(k_x, k_y) exp(i(k_x X + k_y Y + \sqrt{k^2 - k_x^2 - k_y^2 Z})) dk_x dk_y$$

where

$$k = \omega/c$$

with ω being the frequency of operation. Explain how you would evaluate the function $f(k_x, k_y)$ so that the boundary condition

$$F(X, Y, 0) = E(X, Y), (X, Y) \in D$$

is satisfied. Describe an approximate method for evaluating F based on the method of stationary phase, ie, by selecting a contribution to the radiated field pattern from only a single (k_x, k_y) at which the above integrand is stationary w.r.t small variations. Explain why you can make such an approximation.

[5] Consider a parabolic reflector antenna whose surface is defined by the equation

$$\rho^2 = 4az, \rho^2 = x^2 + y^2$$

[a] Prove that its focus is at $(0, 0, a)$ and directrix is the plane $z = -a$ by showing that if (x, y, z) is any point on the parabolic surface, then the distance of this point from the focus equals its distance from the directrix.

[b] Prove that if a ray of light starting from the focus is incident upon the parabolic reflector, then the reflected ray is parallel to the z axis. Do this problem by applying Snell's law of reflection to the light ray, namely [i] the anglse made by the reflected ray and the incident ray with the normal to the surface at the point of incidence are equal and [ii] the incident and reflected rays and the normal fall in one plane.

[c] If an plane electromagnetic wave propagates from the focus to the parabolic surface, then by applying the boundary conditions at the reflected surface assuming no transmitted ray, calculate the electromagnetic field in the reflected ray.

7.3 Article submitted to the Quantum Information Processing Journal for publication

A Model for the refractive index of materials and liquids based on cosmological and quantum mechanical considerations

Abstract

Abstract: We describe a mathematical model based on classical and quantum field theory and statistical field theory for correlating the refractive indices of materials with experiment. The model described in this paper gives a direct approach to computing the RI of a material based on Dirac's quantum mechanics for a system of N interacting particles in an external electromagnetic field taking corrections due to gravitational effects into account. If we solve the Dirac equation using perturbation theory for a single particle in an electromagnetic field we obtain the wave function in the position space representation. We could calculate the quantum averaged electric and magnetic dipole moment of the electron taken with respect to the probability density of the position of the electron described according to Max Born's interpretation of the wave function. This would in turn enable us to determine the permittivity and permeability of the medium in terms of the electric and magnetic fields. However, this analysis does not show how the RI depends upon the temperature of the material. In order to obtain temperature dependence, we consider Dirac's quantum mechanics for an N particle system taking interparticle interactions into account apart from interaction of the particles with an external electromagnetic field. By partial tracing the mixed state Dirac equation over the other particles and then making some approximations we derive a quantum Boltzmann equation for the quantum density operator. This equation is solved using perturbation theory with the initial state as the Gibbs state (which has temperature dependence). The final equilibrium state in the presence of a static electromagnetic field and interparticle interactions will then also depend upon temperature. When this final density matrix is used to compute quantum averages of the electric and magnetic dipole moment, we are able to explain the dependence of the RI on both the electromagnetic field and temperature. The wavelength dependence of the RI can be explained by assuming the background electromagnetic field to be black-body radiation which has the energy density of the electromagnetic field dependent upon both frequency/wavelength and temperature. To this model of the RI, we add cosmological and background gravitational correction effects based on the follwing idea: Gravity affects quantum mechanics via the spinor connection of the gravitational field. This to be introduced into Dirac's equation in order to make it invariant under local Loretnz transformations and arbitrary diffeomorphisms of space-time. Thus, this general relativistic generalization of Dirac's equation gives us the dependence of the wave function on the background metric tensor of curved space-time. We give three independent derivations for the spinor connection of the gravitational field based on standard arguments in gauge field theory and spinor forms of the Riemann curvature tensor. If this background metric is taken to be the Schwarzchild metric, the wave function would depend upon the mass of the blackhole and the gravitational constant while if it is taken to be Robertson-Walker metric for an expanding homogeneous and isotropic universe, then the wave function will also depend on the radius of the universe and hence on Hubble's constant. Calculating the average electric and magnetic dipole moments w.r.t such a wave function would then yield the dependence of the RI on the radius of the expanding universe and on its curvature. By taking fine measurements of the RI, we would then in principle be able to measure Hubble's constant and hence the radius of the universe at the present epoch. It should be noted that using the quantum Boltzmann equation or its gravitationally modified version is based on the quantum theory for a finite number of indistinguishable particles, ie, it is a first quantization approach. If the number of particles is infinite in number, we then have to adopt a second quantization approach based on Fermionic field operators for the Dirac wave function. We explain how to set up such a second quantized Hamiltonian and thereby calculate temperature Green's functions for these by assuming that the state of the second quantized field is given by the Gibbs density with the unperturbed second quantized Hamiltonian. Using this temperature Green's function, we evaluate the average polarization and magnetization of the field as a function of the electromagnetic field, the temperature and the background gravitational field. The final parts of the manuscript focus on deriving the basic cosmological equations for the expanding universe using Newtonian and Eulerian fluid mechanics in terms of the scale factor/radius of the universe and the propagation of inhomogeneities in the matter, temperature and electromagnetic field in such a uniformly expanding universe. The idea is that the perturbations to an initially applied electromagnetic field will depend on the scale factor and temperature and hence if we use classical statistical mechanics to calculate the average electric and magnetic dipole moments in such an electromagnetic field, then the permittivity, permeability and hence the refractive index will also depend upon the scale factor and the temperature. Further, we observe that if the dynamical equations of the expanding universe are quantized (just as Hawking determined the wave function of the radius of the expanding universe) using the Lindlbad open quantum system formalism, then we can in principle calculate the evolving state of the matter velocity, density, temperature and electromagnetic fields using which we can determine the quantum fluctuations in the electric and magnetic dipole moments of a system of charges which would in turn give us the mean square quantum fluctuations in the refractive index. An appendix has been included containing a brief and elementary derivation of the Lindblad master equation for an open quantum system, ie, a system interacting with a bath. The manuscript also contains a short look at what the quantum Boltzmann equation will look like for an open quantum system described by Lindlbad operators.

1. The quantum Boltzmann equation for a plasma

In this section, we derive an approximate nonlinear evolution equation for the density operator of a single particle when the quantum plasma consists of N identical particles interacting with each other and also with an external electromagnetic field. The joint density operator of the N particles satisfies the quantum Liouville or Schrodinger-Von-Neumann equation with the Hamiltonian consisting of a sum of identical Hamiltonians each acting in a single particle Hilbert space plus the sum of identical pairwise interacting potentials of two particles with each one acting in the tensor product of two identical Hilbert spaces. By taking the partial trace of this quantum Liouville equation and making approximations (which in the classical Boltzmann kinetic transport theory are called the molecular chaos approximation), we derive an approximate quadratic nonlinear evolution equation for the single particle density operator in an external electromagnetic field. The single particle Hamiltonians can either be the single particle Schrodinger equation in an external electromagnetic field or a single particle Dirac Hamiltonian or even a single particle Dirac Hamiltonian in curved space-time interacting with an external electromagnetic field. The quadratic nonlinear terms which arise due to the pairwise interaction of particles represent quantum generalization of the so called "collsion term" that appears in the classical Boltzmann equation in kinetic transport theory and which are usually evaluated using classical scattering theory or more specifically using binary elastic collision theory of two particles. It should be noted that our method of deriving the quantum Boltzmann equation by partial tracing is the quantum analogue of the classical BBGKY theory in which one writes down the classical Liouville equation for the distribution function of N particles in phase space (ie, in the joint position-velocity space of all the N particles) and then integrates this equation over the phase space variables of all but the

first particles and then makes the molecular chaos approximation in which the joint distribution of two particles is approximated by a product of the individual distributions.

Suppose that the joint density matrix of N particles is $\rho(123...N)$. It satisfies the Schrodinger equation

$$i\partial_t\rho_t(12...N) = [\sum_{a=1}^N H_a + \sum_{1\leq a<b\leq N} V_{ab}, \rho_t(12..N)]$$

In this equation, if we take a partial traced over $2, 3, ..., N$, we get

$$i\partial_t\rho_{1t} = [H_1, \rho_{1t}] + (N-1)Tr_2[V_{12}, \rho_{12}]$$

and if we take the trace of the same over $3, 4, ..., N$, we get

$$i\partial_t\rho_{12t} = [H_1 + H_2 + V_{12}, \rho_{12t}] + (N-2)Tr_3[V_{13} + V_{23}, \rho_{123t}]$$

We write

$$\rho_{123} = (1/3)(\rho_{12}\otimes\rho_3 + \rho_{13}\otimes\rho_2 + \rho_1\otimes\rho_{23}) + g_{123}$$

where g_{123} is small. Then, neglecting second order of smallness terms like V multiplied with g_{123} gives us the approximate equation

$$i\partial_t\rho_{12t} = [H_1+H_2+V_{12}, \rho_{12}]+((N-2)/3)Tr_3[V_{13}+V_{23}, \rho_{12}\otimes\rho_3+\rho_{13}\otimes\rho_2+\rho_1\otimes\rho_{23}]$$

This is a bit hard to handle. So we content ourselves with the approximation

$$\rho_{12} = \rho_1\otimes\rho_1 + g_{12}$$

where g_{12} is small. We then get approximately,

$$i\partial_t\rho_{1t} = [H_1, \rho_{1t}] + (N-1)Tr_2[V_{12}, \rho_1\otimes\rho_1]$$

Writing

$$V_{12} = \sum_a W_{1a}\otimes W_{2a}$$

gives us

$$Tr_2[V_{12}, \rho_1\otimes\rho_2] = \sum_a Tr(\rho_1 W_{2a})[W_{1a}, \rho_1]$$

and the our Boltzmann equation becomes

$$i\partial_t \rho_1 = [H_1, \rho_1] + (N-1)\sum_a Tr(\rho_1 W_{2a})[W_{1a}, \rho_1]$$

Suppose we make the approximation

$$\rho_{123} = \rho_1 \otimes \rho_1 \otimes \rho_1 + g_{123}$$

where g_{123} is small. Then we get

$$i\partial_t \rho_{12t} = [H_1 + H_2 + V_{12}, \rho_{12}] + (N-2)Tr_3[V_{13} + V_{23}, \rho_1 \otimes \rho_1 \otimes \rho_1]$$

Even this equation is hard to manipulate further without assuming some specific form of the interaction potential V_{12}. We consider

$$\rho_{12} = \rho_1 \otimes \rho_1 + g_{12},$$

$$\rho_{123} = (1/3)(\rho_{12} \otimes \rho_3 + \rho_{13} \otimes \rho_2 + \rho_1 \otimes \rho_{23}) + g_{123}$$

$$= \rho_1 \otimes \rho_1 \otimes \rho_1 + (1/3(g_{12} \otimes \rho_3 + g_{13} \otimes \rho_2 + \rho_1 \otimes g_{23}) + g_{123}$$

We first derive a differential equation for g_{12} after neglecting second order of smallness terms:

$$i\partial_t \rho_{12} = i\partial_t \rho_1 \otimes \rho_1 + i\rho_1 \otimes \partial_t \rho_1$$

$$+ i\partial_t g_{12}$$

$$= [H_1, \rho_1] \otimes \rho_1 + (N-1)Tr_2[V_{12}, \rho_1 \otimes \rho_1] \otimes \rho_1 + \rho_1 \otimes [H_1, \rho_1]$$

$$+ (N-1)\rho_1 \otimes Tr_2[V_{12}, \rho_1 \otimes \rho_1] + i\partial_t g_{12}$$

$$= [H_1 + H_2 + V_{12}, \rho_{12}] + (N-2)Tr_3[V_{12} + V_{13}, \rho_{123}]$$

$$= [H_1 + H_2, \rho_1 \otimes \rho_1] + [V_{12}, \rho_1 \otimes \rho_1]$$

$$+ (N-2)Tr_3[V_{13} + V_{23}, \rho_1 \otimes \rho_1 \otimes \rho_1]$$

After making the appropriate cancellations, we get

$$i\partial_t g_{12} =$$

$$= [V_{12}, \rho_1 \otimes \rho_1] + (N-2)Tr_3[V_{13} + V_{23}, \rho_1 \otimes \rho_1 \otimes \rho_1]$$

$$-(N-1)Tr_2[V_{12}, \rho_1 \otimes \rho_1] \otimes \rho_1$$

$$-(N-1)\rho_1 \times Tr_2[V_{12}, \rho_1 \otimes \rho_1]$$

Note that on writing

$$V_{12} = \sum_a W_{1a} \otimes W_{2a}$$

and using the fact that the $V'_{jk}s$ are identical copies of each other acting on different copies of the tensor product of two identical copies a Hilbert space just as the $H'_k s$ are identical copies of each other acting on different copies of the same Hilbert space, we get

$$Tr_3[V_{13}+V_{23},\rho_1\otimes\rho_1\otimes\rho_1]$$

$$=\sum_a[Tr(\rho_1 W_{2a})([W_{1a},\rho_1]\otimes\rho_1+\rho_1\otimes[W_{1a},\rho_1])]$$

A better approximation to the quantum Boltzmann equation can then be obtained by solving this equation for $g_{12}(t)$ and substituting it into the equation

$$i\partial_t\rho_1=[H_1,\rho_1]+(N-1)Tr_2[V_{12},\rho_{12}]$$

$$=[H_1,\rho_1]+(N-1)Tr_2[V_{12},\rho_1\otimes\rho_1+g_{12}]$$

Formally this equation has the form

$$i\partial_t\rho_1(t)=[H_1,\rho_1(t)]+\delta.F_1(\rho_1(s),s\leq t)$$

where F is an operator valued nonlinear functional of $\rho_1(s), s\leq t$. This equation can be solved upto $O(\delta)$ using first order perturbation theory:

$$\rho_1(t)=U(t)\rho_1(0)U(t)^*+\delta.\int_0^t U(t-\tau)F_\tau(\rho_1(s),s\leq\tau)U(t-\tau)^*d\tau$$

where

$$U(t)=exp(-itH_1)$$

If we consider the Hamiltonian to comprise of an interaction between the particles and an electromagnetic field, then we can write

$$H_1=(p_1+eA(t,r))^2/2m-e\Phi(t,r)\approx p_1^2/2m-e\Phi(t,r)+(e/2m)((p_1,A)+(A,p_1))$$

and the particle interaction potential as

$$V_{12}=V(|r_1-r_2|)$$

In that case, in the position representation, we have

$$[p_1^2,\rho_1]=[p_1,\rho_1].p_1+p_1.[p_1,\rho_1]$$

and noting that p_1 is represented by the kernel

$$p_1(r,r')=-i\nabla_r\delta^3(r-r')=i\nabla'_r\delta^3(r-r')$$

we get

$$[p_1,\rho_1](r,r')=-i\nabla_r\rho_1(r,r')+i\nabla'_r\rho_1(r,r')$$

$$[p_1,\rho_1].p_1(r,r')=\nabla_r.\nabla'_r\rho_1(r,r')-\nabla_r^{'2}\rho_1(r,r')$$

and likewise,

$$p_1.[p_1,\rho_1](r,r')=\nabla_r.\nabla'_r\rho_1(r,r')-\nabla_r^2\rho_1(r,r')$$

$$[(A,p_1),\rho_1](r,r')=[A_k p_{1k},\rho_1](r,r')=$$

$$A_k(t,r)p_{1k}\rho_1(r,r')-\int\rho_1(r,r'')A_k(t,r'')p_{1k}(r'',r')dr''$$

$$= -i(A(t,r), \nabla_r)\rho_1(r,r') - i(\nabla'_r, A(t,r'))\rho_1(r,r'))$$

Therefore with neglect of nonlinear terms in the electromagnetic field, our position space representation of the quantum Boltzmann dynamics of the single particle density operator is given by

$$i\partial_t\rho_1(t,r,r') = (2m)^{-1}(2\nabla_r.\nabla'_r\rho_1(t,r,r') - (\nabla_r^2 + \nabla_r'^2)\rho_1(t,r,r'))$$

$$-(i/m)(A(t,r), \nabla_r)\rho_1(t,r,r') - i(\nabla'_r, A(t,r'))\rho_1(t,r,r')$$

$$-e\Phi(t,r)\rho_1(t,r,r') + e\Phi(t,r')\rho_1(t,r,r')$$

$$+nonlinearterms.$$

A remark on the nonlinear terms:

$$Tr_2[V_{12}\rho_1 \otimes \rho_1](r_1, r_1') =$$

$$V(r_1,r_2)\delta(r_1 - r_1'')\delta(r_2 - r_2')\rho_1(t, r_1'', r_1')\rho_1(t, r_2', r_2)d^3r_2'd^3r_2d$$

$$= \int V(r_1,r_2)\rho_1(t,r_1,r_1')\rho_1(t,r_2,r_2)d^3r_2$$

and likewise,

$$Tr_2[(\rho_1 \otimes \rho_1)V_{12}](r_1, r_1') =$$

$$\int \rho_1(t,r_1,r_1')\rho_1(t,r_2,r_2)V_{12}(r_1',r_2)d^3r_2$$

In principle, using perturbation theory, this quantum Boltzmann equation can be solved for to obtain the single particle density operator kernel $\rho_1(t,r,r')$ as a function of the electromagnetic field and then the average dipole moment of the electron in this electromagnetic field will be given by

$$\mathbf{p}(t) = \int (-e\mathbf{r})\rho_1(t,\mathbf{r},\mathbf{r})d^3r$$

and its average magnetic moment by

$$\mathbf{m}(t) = \int (-e\mathbf{L}(r,r')/2m)\rho_1(t,\mathbf{r}',\mathbf{r})d^3rd^3r'$$

where $\mathbf{L}(r,r')$ is the kernel of the orbital angular momentum operator in the position representation. This would then solve the problem of explaining the origin of permittivity and permeability of the plasma from the quantum statistical mechanical point of view.

2.Perturbative solution of the Boltzmann equation

$$i\partial_t\rho_1(t) = [H_1, \rho_1(t)] + \delta(N-1)Tr_2[V_{12}, \rho_1(t) \otimes \rho_1(t)]$$

where δ is a perturbation parameter. We've also observed that an additional term can be added to this equation to improve its accuracy, namely,

$$\delta^2(N-1)Tr_2[V_{12}, g_{12}]$$

where

$$g_{12}(t) =$$

$$= -i\int_0^t [[V_{12}, \rho_1 \otimes \rho_1] + (N-2)Tr_3[V_{13} + V_{23}, \rho_1 \otimes \rho_1 \otimes \rho_1]$$

$$-(N-1)Tr_2[V_{12}, \rho_1 \otimes \rho_1] \otimes \rho_1$$

$$-(N-1)\rho_1 \times Tr_2[V_{12}, \rho_1 \otimes \rho_1]](s)ds$$

$$= \int_0^t F_2(\rho_1(s) \otimes \rho_1(s), \rho_1(s) \otimes \rho_1(s) \otimes \rho_1(s))ds$$

so our perturbative solution would have the form

$$\rho_1(t) = U(t)\rho_1(0)U(t)^* + \delta.\int_0^t U(t-s)F_1(\rho_1(s) \otimes \rho_1(s))U(t-s)^* ds$$

$$+\delta^2(N-1)\int_0^t U(t-s)[\int_0^s F_2(\rho_1(\tau)\otimes\rho_1(\tau), \rho_1(\tau)\otimes\rho_1(\tau)\otimes\rho_1(\tau))d\tau]U(t-s)^* ds$$

$$= \rho_1(t) = U(t)\rho_1(0)U(t)^* + \delta.\int_0^t U(t-s)F_1(\rho_1(s) \otimes \rho_1(s))U(t-s)^* ds$$

$$+\delta^2(N-1)\int_{0<\tau<s<t} [U(t-s)F_2(\rho_1(\tau)\otimes\rho_1(\tau), \rho_1(\tau)\otimes\rho_1(\tau)\otimes\rho_1(\tau))d\tau]U(t-s)^*]dsd\tau$$

where

$$F_1(\rho_1 \otimes \rho_1) = (N-1)Tr_2[V_{12}, \rho_1 \otimes \rho_1]$$

3.The quantum Boltzmann equation derived from the Dirac relativistic wave equation

$$H(t) = (\alpha, -i\nabla + eA) + \beta m - e\Phi = H_0 + H_I(t)$$

where

$$H_0 = (\alpha, -i\nabla) + \beta m, H_I(t) = e(\alpha, A) - e\Phi$$

Let $V_{12} = V(r_1, r_2)$ be the interaction potential between two particles. Our aim is to formulate the Quantum Boltzmann Equation

$$i\partial_t\rho(t) = [H(t), \rho(t)] + (N-1)Tr_2[V_{12}, \rho(t) \otimes \rho(t)]$$

in the position space representation. We first note that $\rho(t, r, r')$ for each t, r, r' is a 4×4 matrix and

$$[H_0, \rho] = [\alpha_k p_k + \beta m, \rho](r, r')$$

$$= \alpha_k(-i\partial_k \rho(t,r,r')) - i\partial'_k \rho(t,r,r')\alpha_k$$

and

$$[\beta,\rho](r,r') = \beta.\rho(t,r,r') - \rho(t,r,r')\beta$$

Further,

$$((\alpha,A)\rho)(r,r') = A_k(t,r)\alpha_k\rho(t,r,r') = (\alpha,A(t,r))\rho(t,r,r')$$

$$(\rho(\alpha,A))(r,r') = \rho(t,r,r')(\alpha,A(t,r'))$$

Thus,

$$[(\alpha,A),\rho](r,r') = (\alpha,A(t,r))\rho(t,r,r') - \rho(t,r,r')(\alpha,A(t,r'))$$

and likewise,

$$[\Phi,\rho](r,r') = (\Phi(t,r) - \Phi(t,r'))\rho(t,r,r')$$

Remark: The cumulative distribution function of an observable X in the state ρ is given by

$$F_\rho(x) = Tr[\rho.\chi_{(-\infty,x]}(X)]$$

where $\chi_E(x)$ is the indicator function of the set $E \subset \mathbb{R}$ and this formula can be applied to calculate the average electric and magnetic dipole moments of the electron in the evolving Boltzmann state.

4. Quantum statistical field theory

Rather than using the quantum Boltzmann equation to calculate the average polarization and magnetization of a medium in an external electromagnetic field, we can use quantum statistical field theory based on second quantization principles of the wave function operator field. The quantum Boltzmann equation is based on first quantized quantum mechanics for a system comprising of a finite number of particles. However, quantum statistical field theory does not put any restriction on the number of particles. The number of particles here is assumed to be infinite and therefore its wave function is an operator valued Fermionic field. Such a Fermionic wave function field operator can equivalently be looked upon as being generated by a countably infinite number of Fermionic creation and annihilation operators modulated by a by a sequence of orthonormal stationary state ordinary wave functions.

Let $\psi(t,r)$ denote the wave operator field of second quantized matter. The second quantized Hamiltonian in the Dirac picture is given by

$$H = \int \psi(t,r)^*((\alpha,-i\nabla+eA(r))+\beta m)\psi(t,r)d^3r + \int V_{\mu\nu}(r,r')\psi(t,r)^*\alpha^\mu\psi(t,r)\psi(t,r')^*\alpha^\nu\psi(t,r')$$

just as in the Hartree-Fock theory. The wave operator fields satisfy the canonical equal time anticommutation relations

$$\{\psi_l(t,r),\psi_m(t,r')^*\} = \delta_{lm}\delta^3(r-r')$$

and the second term in the second quantized Hamiltonian represents the interaction between Dirac charges and currents at two different spatial points. The wave operator fields $\psi(t,r)$ evolve according to the Heisenberg dynamics

$$\partial_t \psi(t,r) = i[H, \psi(t,r)]$$

The electronic polarization operator field is given by

$$P(t,r) = -e\mathbf{r}\psi(t,r)^* \psi(t,r)$$

This is the dipole moment operator per unit volume. Let ρ_0 denote the initial state of the quantum system, say the Gibbs state:

$$\rho_0 = exp(-\beta H)/Z(\beta), Z(\beta) = Tr(exp(-\beta H)), \beta = 1/kT$$

Since we are adopting the Heisenberg picture, this state does not evolve with time, only the observables evolve with time. The average Polarization of the medium at time t is therefore given by

$$< P > (t,r) = Tr(\rho_0.P(t,r))$$

The magnetic dipole moment operator field per unit volume is given by

$$M(t,r) = \psi(t,r)^*(-e(\mathbf{L} + g\sigma)/2m)\psi(t,r)$$

and its average value is given by

$$< M > (t,r) = Tr(\rho_0 M(t,r))$$

To calculate these averages, we must first determine the dynamics of the temperature Green's function

$$G(t,r|t',r') = Tr(\rho_0 T\{\psi(t,r)\psi(t',r')^*\})$$

where T is the time ordering operator. Note that since the magnetic vector potential is not assumed to vary with time, the total Hamiltonian operator is a constant of the motion and hence so is the density operator ρ_0. Note that from the canonical anticommutation rules,

$$[\psi(t,r')^* \alpha^\mu \psi(t,r'), \psi_k(t,r)] =$$

$$= [\alpha^\mu(l,m)\psi_l(t,r')^* \psi_m(t,r'), \psi_k(t,r)] =$$

$$-\alpha^\mu(l,m)\delta_{lk}\delta^3(r'-r)\psi_m(t,r')$$

$$= -\delta^3(r-r')\alpha^\mu(k,m)\psi_m(t,r) = -\delta^3(r-r')[\alpha^\mu \psi(t,r)]_k$$

Equivalently, in vector notation,

$$[\psi(t,r')^* \alpha^\mu \psi(t,r'), \psi(t,r)] = -\delta^3(r-r')\alpha^\mu \psi(t,r)$$

It follows that if we define

$$J^\mu(t,r) = \psi(t,r)^*\alpha^\mu\psi(t,r)$$

then

$$[J^\mu(t,r'),\psi(t,r)] = -\delta^3(r-r')\alpha^\mu\psi(t,r)$$

and therefore,

$$[J^\mu(t,r')J^\nu(t,r''),\psi(t,r)] =$$
$$-[\delta^3(r-r'')J^\mu(t,r')\alpha^\nu\psi(t,r) + \delta^3(r'-r)\alpha^\mu\psi(t,r)J^\nu(t,r'')]$$

Then,

$$[\int V_{\mu\nu}(r',r'')J^\mu(r')J^\nu(r'')d^3r'd^3r'',\psi(t,r)] =$$
$$-[\int V_{\mu\nu}(r',r)J^\mu(t,r')d^3r']\alpha^\nu\psi(t,r)$$
$$-\int \alpha^\mu\psi(t,r)[\int V_{\mu\nu}(r,r')J^\nu(t,r')d^3r']$$

We may assume without loss of generality that

$$V_{\mu\nu}(r,r') = V_{\nu\mu}(r',r)$$

and then deduce that

$$[\int V_{\mu\nu}(r',r'')J^\mu(r')J^\nu(r'')d^3r'd^3r'',\psi(t,r)] =$$
$$= -\{\alpha^\mu\psi(t,r),\int V_{\mu\nu}(r,r')J^\nu(t,r')d^3r'\}$$

We therefore obtain from the Heisenberg dynamics the following dynamical equation for the wave field operator $\psi(t,r)$.

$$i\partial_t\psi(t,r) = H_{D0}\psi(t,r) + \{\alpha^\mu\psi(t,r),\int V_{\mu\nu}(r,r')J^\nu(t,r')d^3r'\}$$

where

$$H_{D0} = (\alpha,-i\nabla) + \beta m$$

is the first quantized free particle Dirac Hamiltonian. We can further approximate this equation by replacing $J^\mu(t,r')$ on the rhs by its quantum average

$$< J^\mu(t,r) >= Tr(\rho_0(T)J^\mu(t,r))$$

It should be noted that this average current density can be expressed in terms of the temperature Green's function as

$$< J^\mu(t,r) >= -Tr[\alpha^\mu G(t,r|t,r')]|_{r'\to r}$$

where the trace here is an ordinary matrix trace for 4×4 matrices.

Reference:Fetter and Walecka, "Quantum Theory of Many Particle Systems", Dover, 1971.

5. Relating the refractive index of a material to the metric tensor of space-time

[a]**Brief outline of the approach**

The curvature of space-time affects quantum phenomena. For example, in order to take into account the space-time curvature, we have to write down Dirac's equation in curved space-time and then formulate the quantum Boltzmann equation by starting from such a generalized Dirac equation. This is accomplished by assuming that the spinor connection of the gravitational field is chosen so that the Dirac equation based on such covariant derivative remains invariant under local Lorentz transformations. If such gravitational considerations are to be incorporated into the quantum Boltzmann equation, then one must write down Dirac's equation in curved space-time in Hamiltonian form and identify the extra terms in the Hamiltonian of each particle that come due to its interaction with gravity ie those extra terms will involve the tetrad basis and the spinor connection of the gravitational field. The quantum Boltzmann equation will have the same structure but with these additional terms in the single particle Hamiltonian involving the gravitational field. Likewise, if one has to incorporate these gravitational terms in the temperature Green's function based on statistical quantum field theory, then again one must replace the second quantized Hamiltonian term with $\int \psi(t,r)^* H\psi(t,r)d^3r$ where

$$H = (\alpha, -\nabla + eA) + \beta m + \delta H_g$$

where δH_g is the correction term in the Dirac Hamiltonian coming from the tetrad and spinor connection terms of the gravitational field.

The main computations

Let V_a^μ be a tetrad basis for our curved space-time and let $\Gamma_\mu = \Gamma^\mu_{ab}[\gamma^a, \gamma^b]$ denote the spinor connection of the gravitational field. Then, the four component wave function satisfies

$$[V_a^\mu \gamma^a(i\partial_\mu + eA_\mu + i\Gamma_\mu) - m]\psi(x) = 0$$

From this equation, we can infer what the generalized Dirac Hamiltonian must be. This is achieved by separating the time derivative component from the spatial derivative components:

$$iV_a^0\gamma^a\partial_0\psi + [iV_a^r\gamma^a\partial_r + eV_a^\mu\gamma^a A_\mu + eV_a^\mu\gamma^a A_\mu + iV_a^\mu\gamma^a\Gamma_\mu - V_a^\mu\gamma^a m]\psi = 0$$

Now multiplying both sides of this equation by $V_b^0\gamma^b$ and using

$$V_a^0 V_b^a\gamma^a\gamma^b = (1/2)\eta^{ab}V_a^0 V_b^0 = (1/2)g^{00}$$

where η^{ab} is the Minkowski metric of flat space-time and $g^{\mu\nu}$ is the exact contravariant metric of our curved space-time, we get

$$ig^{00}\partial_0\psi+[iV_b^0V_a^r\gamma^b\gamma^a\partial_r+eV_b^0V_a^\mu\gamma^b\gamma^aA_\mu+iV_b^0V_a^\mu\gamma^b\gamma^a\Gamma_\mu-V_b^0V_a^\mu\gamma^b\gamma^am]\psi=0$$

This equation can be expressed in the standard Hamiltonian form by defining the curved space time Dirac Hamiltonian in an electromagnetic field as

$$H=(g^{00})^{-1}V_b^0V_a^r\gamma^b\gamma^a(-i\partial_r)-(g^{00})^{-1}(eV_b^0V_a^\mu\gamma^b\gamma^aA_\mu)+(g^{00})^{-1}V_b^0V_a^\mu\gamma^b\gamma^a(-i\Gamma_\mu)$$
$$+(g^{00})^{-1}V_b^0V_a^\mu\gamma^b\gamma^am]$$

As an example of this calculation, consider the Schwarzchild metric in which

$$g_{00}=\alpha(r)=1-2m/r,g_1=-\alpha(r)^{-1},g_{22}=-r^2,g_{33}=-r^2sin^2(\theta)$$

We have

$$d\tau^2=g_{\mu\nu}dx^\mu dx^\nu=(\omega_0)^2-\omega_1^2-\omega_2^2-\omega_3^2$$

where

$$\omega_0=\sqrt{\alpha(r)}dt,\omega_1=\sqrt{\alpha(r)^{-1}}dr,$$
$$\omega_2=rd\theta,\omega_3=rsin(\theta)d\phi$$

Thus, since

$$g^{\mu\nu}=\eta^{ab}V_a^\mu V_b^\nu,g_{\mu\nu}=\eta_{ab}V_\mu^aV_\nu^b$$

we get

$$d\tau^2=\eta_{ab}V_\mu^aV_\nu^bdx^\mu dx^\nu$$
$$=(V_\mu^0dx^\mu)^2-(V_\mu^1dx^\mu)^2-(V_\mu^2dx^\mu)^2-(V_\mu^3dx^\mu)^2$$

Thus,

$$\omega_0=\sqrt{\alpha(r)}dt=V_\mu^0dx^\mu=V_0^0dt+V_1^0dr+V_2^0d\theta+V_3^0d\phi,$$

so that

$$\sqrt{\alpha(r)}=V_0^0,V_1^0=V_2^0=V_3^0=0,$$

Note that in these conventions,

$$(V_\mu^0)=(V_0^0,V_1^0,V_2^0,V_3^0)$$

Likewise,

$$\omega_1=V_\mu^1dx^\mu=\alpha(r)^{-1/2}dr$$

and therefore,

$$V_0^1=0,V_1^1=\alpha(r)^{-1/2},V_2^1=0,V_3^1=0$$

Note that

$$(V_\mu^1)=(V_0^1,V_1^1,V_2^1,V_3^1)$$
$$\omega_2=V_\mu^2dx^\mu=rd\theta,\omega_3=V_\mu^3dx^\mu=r.sin(\theta)d\phi$$

so that

$$V_0^2 = 0, V_1^2 = 0, V_2^2 = r, V_3^2 = 0$$

$$V_0^3 = 0, V_1^3 = 0, V_2^3 = 0, V_3^3 = r.sin(\theta)$$

The spinor connection of the gravitational field is given by

$$\Gamma_\mu = V_{a\nu} V^\nu_{b:\mu}[\gamma^a, \gamma^b] = \omega_\mu^{ab}[\gamma^a, \gamma^b]$$

This can be derived in various ways, one by using the fact that the covariant derivative of the tetrad is zero, ie

$$V^a_{\mu,\nu} - \Gamma^\rho_{\mu\nu} V^a_\rho + \omega_\nu^{ab} V_{b\mu} = 0$$

or equivalently,

$$V^a_{\mu:\nu} + \omega_\nu^{ab} V_{b\mu} = 0$$

The second derivation is based on starting with the Einstein-Hilbert Lagrangian for the gravitational field, namely the curvature tensor in spinor notation:

$$R = R^{ab}_{\mu\nu} V^\mu_a V^\nu_b,$$

$$R^{ab}_{\mu\nu} = \omega^{ab}_{\nu,\mu} - \omega^{ab}_{\mu,\nu} - [\omega_\mu, \omega_\nu]^{ab}$$

and setting the variational derivative of $\int R d^4x$ w.r.t ω_μ^{ab} to zero to arrive at an algebraic equation for ω_μ^{ab} which when solved yields the desired form of the gravitational spinor connection. The third and most interesting way to derive the form of the spinor connection of the gravitational field is based on the use of non-Abelian gauge group theory with the gauge group now being the spinor representation of the Lorentz group. In other words, the transformation law of the spinor connection of the gravitational field under a local Lorentz transformation should be such that the Dirac equation remains invariant under it. Let $\Gamma_\mu(x)$ denote the spinor connection in one frame. The corresponding covariant derivative is $\nabla_\mu = \partial_\mu + \Gamma_\mu$. Now suppose we apply a local Lorentz transformation $\Lambda(x)$. Then the Dirac wave function transforms from $\psi(x)$ to $D(\Lambda(x))\psi(x)$ and for the Dirac equation to remain invariant under this transformation, we require that $\Gamma_\mu(x)$ should transform to $\Gamma'_\mu(x)$ where

$$D(\Lambda(x))(\partial_\mu + \Gamma_\mu(x))D(\Lambda(x))^{-1} =$$

$$\partial_\mu + \Gamma'_\mu(x)$$

This is equivalent to requiring that

$$\Gamma'_\mu(x) = D(\Lambda(x))\Gamma_\mu(x)D(\Lambda(x))^{-1} + D(\Lambda(x))(\partial_\mu D(\Lambda(x))^{-1})$$

or equivalently that

$$\Gamma'_\mu(x) = D(\Lambda(x))\Gamma_\mu(x)D(\Lambda(x))^{-1} - (\partial_\mu D(\Lambda(x)))D(\Lambda(x))^{-1}$$

Equivalently, if

$$\Lambda(x) = I + \omega(x)$$

is an infinitesimal local Lorentz transformation, then we require that

$$\delta\Gamma_\mu(x) = \Gamma'_\mu(x) - \Gamma_\mu(x) =$$

$$[D(\omega(x)), \Gamma_\mu(x)] + \partial_\mu D(\omega(x))$$

where

$$D(\omega(x)) = (1/4)[\gamma^a, \gamma^b]\omega_{ab}(x)$$

is the differential of the spinor representation (interpreted in terms of Lie algebra representations) of the Lorentz group evaluated at $\omega(x)$. On the other hand, if we take $\Gamma_\mu(x) = V_{a\nu:\mu} V_b^\nu [\gamma^a, \gamma^b]/2$, then under this infinitesimal local Lorentz transformation, it changes by

$$\delta\Gamma_\mu(x) = ([\gamma^a, \gamma^b]/4)[(\delta V_{a\nu:\mu})V_b^\nu + V_{a\nu:\mu}\delta V_b^\nu]$$

where

$$\delta V_{a\nu} = \omega_a^b(x) V_{b\nu}$$

so that

$$\delta V_{a\nu:\mu} = \omega_a^b V_{b\nu:\mu} + \omega_{a,\mu}^b V_{b\nu}$$

since the ω_b^a's and ω_{ab}'s transform as scalar fields under diffeomorphisms of space-time. By comparing these two transformation laws, we can show that they are identical using the canonical commutation relations for the Loretnz algebra generators.

6.Relating the refractive index to Lindblad noise operators when the universe as a system interacts with a bath

Let H be the Hamiltonian of an open quantum system consisting of N identical particles having a density operator $\rho(t) = \rho_{12...N}(t)$. Its dynamical equations are

$$i\rho'(t) = [H, \rho(t)] + \theta(\rho(t))$$

where

$$\theta(\rho) = (-1/2)\sum_{k=1}^{p}[L_k^* L_k \rho(t) + \rho(t) L_k^* L_k - 2L_k \rho(t) L_k^*]$$

$L_k, k = 1, 2, ..., p$ are the Lindblad operators. We wish to derive a quantum Boltzmann equation from this by partial tracing out over all but the first particle Hilbert space. As usual, let

$$H = \sum_{a=1}^{N} H_a + \sum_{1 \le a < b \le N} V_{ab}$$

Then we get on partial tracing,

$$i\rho_1'(t) = [H_1, \rho_1(t)] + (N-1)Tr[V_{12}, \rho_1(t) \otimes \rho_1(t)]$$

$$+Tr_{23...N}(\theta(\rho(t))$$

Since the particles are assumed to have identical interactions with the bath, The Lindblad operators can be expressed as

$$L_k = \sum_{i_1,...,i_N} c_k(i_1,...,i_N)L_{k,i_1} \otimes ... \otimes L_{k,i_N}$$

where c_k is permutation invariant. Thus, we can even write

$$L_k = \sum_{i=1}^{q} L_{k,i}^{\otimes N}$$

and then observe that

$$L_k^* L_k = \sum_{i,j=1}^{q} (L_{k,i}^* L_{k,j})^{\otimes N}$$

and

$$Tr_{23...N}(L_k^* L_k \rho(t)) \approx \sum_{i,j} (L_{k,i}^* L_{k,j} \rho_1(t))(Tr(L_{k,i}^* L_{k,j} \rho_1(t)))^{N-1}$$

$$Tr_{23..N}(L_k \rho(t) L_k^*) \approx L_{k,i} \rho_1(t) L_{k,j}^* (Tr(L_{k,i} \rho_1(t) L_{k,j}^*))^{N-1}$$

and

$$Tr_{23...N}(\rho(t) L_k^* L_k) \approx \sum_{i,j} (\rho_1(t) L_{k,i}^* L_{k,j})(Tr(\rho_1(t) L_{k,i}^* L_{k,j}))^{N-1}$$

Thus our approximate quantum Boltzmann equation for an open quantum system for N particles contains polynomial terms in the one particle densities.

7. Heat and mass transfer equations in fluid dynamics with cosmological applications $(1/2)v^2$ is the fluid kinetic energy per unit mass. Let ϵ denote the internal energy of the fluid per unit mass. Then if s denotes the entropy per unit mass of the fluid, we have the basic first law of thermodynamics:

$$Tds = d\epsilon + pd(1/\rho)$$

and if w denotes the fluid enthalpy per unit mass, then

$$w = \epsilon + p/\rho$$

Thus,

$$dw = d\epsilon + pd(1/\rho) + dp/\rho = Tds + dp/\rho$$

In particular, if no heat is being pumped into the fluid, then $ds = 0$ and the above equation reduces to

$$dw = dp/\rho$$

The total energy per unit mass of the fluid is

$$u = \epsilon + v^2/2$$

and $\rho.u$ is the fluid energy per unit volume. We use the above first law of thermodynamics along with the fluid momentum equation

$$\rho dv/dt = \rho((v,\nabla)v + v_{,t}) = -\nabla p - \rho\nabla\Phi + div(\sigma)$$

and the equation of continuity

$$\rho_{,t} + div(\rho v) = 0$$

to derive the energy equation, ie, the differential equation satisfied by u.

Remark: Let σ_{ij} denote the stress tensor of fluid. The rate at which is stress does work on the fluid or equivalently, the rate at which heat is dissipated in the fluid due to frictional/viscous and thermal effects is given by

$$\int_S n_i\sigma_{ij}v_j dS = \int_V (\sigma_{ij}v_j)_{,i} d^3r$$

where V is the volume of the fluid and $S = \partial V$ is the surface of the fluid that bounds the volume V. Thus, the rate of dissipation of heat in the fluid per unit volume is given by

$$(\sigma_{ij}v_j)_{,i} = div(\sigma.v)$$

This dissipation of heat contributes to a decrease in the fluid energy. Thus, the energy equation of the fluid can be expressed as

$$\partial(\rho u)/\partial t + div(\rho w\mathbf{v}) = div(\sigma.\mathbf{v}) + \nabla.(D\nabla T)$$

where

$$w = u + p/\rho$$

is the fluid enthalpy per unit mass. Note that w describes the fluid internal energy plus its kinetic energy plus its pressure energy and therefore the energy flux in the fluid should be taken as $\rho w\mathbf{v}$ rather than ρuv. This is in accordance with Bernoulli's equation which states that in the absence of external forces, the enthalpy per unit mass $u + p/\rho$ is a constant along a streamline.

It should be noted that the viscous effects on the dynamics of the fluid are completely contained in the stress tensor which means that the momentum equation of the fluid should be taken as

$$\rho(v_{,t} + (v,\nabla)v) = -\nabla p - \rho\nabla\Phi + div(\sigma)$$

where

$$(div\sigma)_i = \sigma_{ij,j}$$

is the stress force per unit volume of the fluid. Note that the stress force on a volume V of the fluid bounded by a surface S is given by

$$\int_S \sigma_{ij}n_j dS = \int_V \sigma_{ij,j} d^3r$$

and hence $\sigma_{ij,j}$ is to be interpreted as the ith component of the stress force on the fluid per unit volume.

Remark: $div(\sigma.\mathbf{v})$ equals the rate at which energy is pumped into a unit volume of the the fluid by the surrounding fluid layers due to viscous friction while $\mathbf{q} = -D\nabla T$ is the heat flux caused by temperature gradients. Therefore, $-div\mathbf{q} = div(D\nabla T)$ equals the rate at which heat flows into unit volume of the fluid from the surrounding fluid layers due to temperature gradients. $\int_S(-\hat{n}).\sigma.(-\mathbf{v})dS = \int_V div(\sigma.\mathbf{v})d^3r$ is the rate at which thermal energy is pumped into the volume V bounded by the surface S due to friction between the fluid layers just outside S with the layers just within S. Note that the stress tensor σ which is the frictional force per unit area acts just outside S where the fluid velocity is $\mathbf{v}$ and hence the velocity of the fluid just inside S is $-\mathbf{v}$ relative to that just outside, the rate at which the ith component of the viscous forces from outside do work per unit area on the fluid layers just within S having unit normal along the jth coordinate direction is $(-v_i\sigma_{ij})$ and since this force per unit area acts from the outside of S towards the inside of S, the rate at which these forces do work on a unit area of the surface having outward normal $\mathbf{n}$ and hence unit inward normal $-\mathbf{n}$ is given by $(-v_i\sigma_{ij}(-n_j)) = (-\mathbf{n}).\sigma.(-\mathbf{v})$. A better way to say the same thing is as follows: σ_{ij} is the ith component of the viscous stress force acting per unit area of the fluid having normal along the jth direction within the fluid just inside S on the layer just outside. So $-\sigma_{ij}$ is the stress force per unit area exerted by the layers just outside on the layers just inside. Thus the rate at which these outside layers to work on the layers just within is $(v_i(-\sigma_{ij})(-n_j) = v.\sigma.n = n.\sigma.v$. Thus the total rate at which the viscous forces do work on the volume of the fluid within V bounded by the surface S equals $\int n.\sigma.vd^3r = \int_V div(\sigma.v)d^3r$.

In studying heat transfer within a fluid, our aim is to obtain an equations for the rate of change of the entropy with time per unit mass of the fluid. We compute

$$T.ds/dt = d\epsilon/dt - p\rho^{-2}d\rho/dt$$

$$d/dt(\rho u) = d/dt(\rho(\epsilon + v^2/2))$$

$$(\epsilon + v^2/2)d\rho/dt + \rho(d\epsilon/dt + (v, dv/dt))$$

since

$$d\rho/dt = \partial\rho/\partial t + (v, \nabla)\rho = -\rho.divv$$

by the equation of continuity,

$$d(\rho u)/dt = -\rho udivv + \rho(Tds/dt - p.divv + \rho(v, dv/dt)$$

$$= -\rho wdivv + \rho Tds/dt + (v, \rho.dv/dt)$$

where

$$w = u + p/\rho$$

is the enthalpy per unit mass of the fluid. Substituting into this the momentum equation for $\rho.dv/dt$ gives us

$$d(\rho u)/dt = -\rho wdivv + \rho.Tds/dt + (v, -\nabla p + F + div\sigma)$$

where F is the external force per unit volume. Using the energy equation

$$\partial(\rho u)/\partial t + div(\rho w\mathbf{v}) = div(\sigma.\mathbf{v}) + \nabla.(D\nabla T)$$

and observing that

$$div(\rho w\mathbf{v}) = div(\rho u\mathbf{v}) + div(p\mathbf{v})$$

so that the energy equation can be rewritten as

$$d(\rho u)/dt + \rho u.div(\mathbf{v}) + div(p\mathbf{v}) = div(\sigma.\mathbf{v}) + \nabla.(D\nabla T) + (F, v)$$

in this gives us

$$div(\sigma.\mathbf{v})+\nabla.(D\nabla T)-div(p\mathbf{v})-\rho u div\mathbf{v} = -\rho w div v+\rho.T.ds/dt+(v, -\nabla p+div\sigma)$$

Now

$$div(\sigma.v) = (\sigma_{ij}v_j)_{,i} = \sigma_{ij}v_{j,i} + \sigma_{ij,i}v_j$$

and

$$(v, div\sigma) = v_i\sigma_{ij,j} = v_j\sigma_{ij,i}$$

by the symmetry of the stress tensor. Thus, we get after making a cancellation,

$$\sigma_{ij}v_{j,i} + \nabla.(D\nabla T) - div(p\mathbf{v}) - \rho u div v + \rho w div v - \rho T ds/dt + (v, \nabla p) = 0$$

Now,

$$\begin{aligned}\rho w.div v + (v, \nabla p) &= (\rho u + p)div v + div(pv) - p.div v \\ &= \rho u.div v + div(pv)\end{aligned}$$

and hence we get

$$\rho.T ds/dt = \sigma_{ij}v_{j,i} + \nabla.(D\nabla T)$$

Writing

$$T.ds/dt = C_p dT/dt$$

where C_p is the specific heat at constant pressure

$$C_p = T.\partial s/\partial T|_p$$

we can express the above heat transfer equation in the form

$$\rho C_p dT/dt = \sigma_{ij}v_{j,i} + \nabla.(D\nabla T)$$

which is an equation that describes both heat convection and heat diffusion.

8. Cosmological considerations

Consider the Newtonian model for cosmic expansion: Energy/mass conservation for matter within the entire sphere gives us the equation

$$d(4\pi\rho S^3c^2/3)/dt = 0$$

or equivalently,

$$d(\rho S^3)/dt = 0$$

provided that we neglect the pressure contribution to the energy. The energy conservation equation of a point mass on the surface of sphere gives

$$S'^2/2 - GM/S = k, M = 4\pi\rho S^3/3$$

This second equation can be expressed as

$$S'^2/2 - 4\pi\rho S^2/3 = k$$

Let us now take pressure into consideration and formulate our equations. A sphere having comoving radius r contains matter of mass $4\pi\rho(Sr)^3/3$ and the rate at which the energy of this sphere increases with time equals the total rate at which pressure forces from outside it do work on it. This rate is clearly $-p.4\pi(Sr)^2(S'r)$. Thus we obtain the equation of motion

$$d/dt(4\pi\rho(Sr)^3c^2/3) = -p.4\pi(Sr)^2S'r$$

or equivalently,

$$d(\rho S^3)/dt + 3pS^2S'/c^2 = 0$$

using Einstein's energy-mass relationship of special relativity. In the limit when $p/c^2 << \rho$, this equation reduces to the previously obtained equation of matter conservation. Suppose we wish to quantize these dynamical equations. Let us first neglect the pressure so that we are left with just a single equation

$$S'^2/2 - GM/S = k$$

with M a constant. The Hamiltonian for this system with S as the canonical position coordinate and P as the canonical momentum coordinate is given by

$$H(S,P) = P^2/2 - GM/S$$

and if we write down the equation of the evolution operator $U(t)$ taking bath noise outside our universe into account, then we get the HP-qsde

$$dU(t) = [-(iH + P)dt + LdA(t) - L^*dA(t)^*]U(t), P = LL^*/2$$

More precisely, we assume that $S(t)$ is a classical solution to the expansion dynamics and $\delta S(t) = q(t)$ is a small quantum fluctuation. Then the slowly time varying Hamiltonian is given by

$$H(t,q,\pi) = \pi^2/2 - GM/(S(t) + q)$$

where π the canonical momentum is given by

$$\pi = S'(t) + q'$$

In the absence of noise, the wave function of q $\psi(t,q)$ therefore satisfies the Schrodinger equation

$$i\partial_t\psi(t,q) = -\partial^2\psi(t,q)/\partial q^2 - (GM/(S(t)+q))\psi(t,q)$$

and this is one way of determining the wave function of the expanding universe.

How to calculate the permittivity and permeability in an expanding universe: Let

$$S_{\mu\nu} = (1/4)F_{\alpha\beta}F^{\alpha\beta}g_{\mu\nu} - F_{\mu\alpha}F^{\alpha}_{\nu}$$

denote the energy-momentum tensor of the electromagnetic field. This is a random quantity. The average value of the energy density S_{00} divided by 3 is known to be the pressure $p(t)$ if it is isotropic. We note that in special relativity,

$$S_{00} = (1/4)(-2F_{0r}^2 + F_{rs}^2) + F_{0r}^2 = (1/2)(F_{0r}^2 + F_{rs}^2/2) = (1/2)(E^2+B^2)$$

which is the energy density and further,

$$F_{\alpha\beta}F^{\alpha\beta} = -2F_{0r}^2 + F_{rs}^2 = -2(E^2-B^2)$$

so that $(-1/4)F_{\alpha\beta}F^{\alpha\beta} = (1/2)(E^2-B^2)$ is the Lagrangian density of the field.

$$S_{rs} = (1/2)(E^2-B^2)\delta_{rs} - (F_{r0}F_{s0} - F_{rk}F_{sk})$$

Thus,

$$S_{11} = (1/2)(E^2-B^2) - (E_1^2 - B_2^2 - B_3^2)$$

and likewise for S_{22} and S_{33}. It is clear that in the case of isotropic radiation fields,

$$< S_{00} >= U =< E^2+B^2)/2 >$$

is the average energy density of the field and

$$< S_{11} >=< S_{22} >=< S_{33} >= U/3$$

since for isotropic radiation fields (as in the case of black-body radiation),

$$< E_k^2 >=< B_k >^2 = U/3, k = 1,2,3$$

It is for this reason, that when we consider the expanding universe along with the cosmic microwave background radiation that it encloses, in the energy momentum tensor of the matter field

$$\rho v^\mu v^\nu - pg^{\mu\nu}$$

we replace the second term, ie, the pressure term by the energy-momentum tensor of the radiation field because radiation accounts for the pressure The total energy-momentum tensor of the matter plus radiation field is therefore taken as

$$T^{\mu\nu} = \rho v^{\mu} v^{\nu} + S^{\mu\nu}$$

where

$$S_{\mu\nu} = (1/4) F_{\alpha\beta} F^{\alpha\beta} g_{\mu\nu} - F_{\mu\alpha} F_{\nu}^{\alpha}$$

is the energy-momentum tensor of the radiation field. Now suppose we have a single electric dipole $\mathbf{p}$ that interacts with a slowly time varying electric field. Suppose that this this dipole also has a magnetic moment $\mathbf{m}$ that interacts with a slowly time varying magnetic field. For example, an atom or an ion or a molecule or an ionized molecule can be regarded as having both an electric and a magnetic dipole moment. The interaction energy is given by

$$E = -\mathbf{p}.\mathbf{E}(t) - \mathbf{m}.\mathbf{B}(t)$$

and this is a function of the angles θ_1, θ_2 between $\mathbf{p}$ and $\mathbf{E}(t)$ as well as between $\mathbf{m}$ and $\mathbf{B}(t)$. If there is an apriori relationship between θ_1 and θ_2 defined by a measure $d\mu(\theta_1, \theta_2)$, then the average electric and magnetic dipole moments in this slowly time varying electric and magnetic fields are given by

$$< \mathbf{p} > (\mathbf{E}(t), \mathbf{B}(t)) = \frac{exp(\beta(\mathbf{p}.\mathbf{E}(t) + \mathbf{m}.\mathbf{B}(t))\mathbf{p} d\mu(, \mathbf{m})}{Z(\beta)},$$

$$< \mathbf{m} > (\mathbf{E}(t), \mathbf{B}(t)) = \frac{exp(\beta(\mathbf{p}.\mathbf{E}(t) + \mathbf{m}.\mathbf{B}(t))\mathbf{m} d\mu(, \mathbf{m})}{Z(\beta)},$$

where

$$Z(\beta) = \int exp(\beta(\mathbf{p}.\mathbf{E}(t) + \mathbf{m}.\mathbf{B}(t))) d\mu(\mathbf{p}, \mathbf{m})$$

is the classical partition function. It should be noted that while solving the Maxwell equations in a background gravitational field corresponding to the expanding universe, we take as our unperturbed fields $\mathbf{E}_0(t, \mathbf{r}), \mathbf{B}_0(t, \mathbf{r})$ and using perturbation theory with the metric perturbations from flat space-time being regarded as first order of smallness quantities, or equivalently in the cosmic expansion scenario, the scale factor of the expanding universe minus one $\delta S(t)$ as being of the first order of smallness, we solve the Maxwell equations to express the first order change in the electric and magnetic fields in terms of the unperturbed fields and the scale factor. These gravitationally perturbed fields we denote by

$$\mathbf{E}(t, \mathbf{r}) = \mathbf{E}_0(t, \mathbf{r}) + \delta\mathbf{E}(t, \mathbf{r}, \delta S)$$

$$\mathbf{B}(t, \mathbf{r}) = \mathbf{B}_0(t, \mathbf{r}) + \delta\mathbf{B}(t, \mathbf{r}, \delta S)$$

and it is precisely these perturbed fields which we use to calculate the interaction energy between the electric and magnetic dipoles and the fields. It follows immediately, that the average electric and magnetic dipole moment per unit volume at time t will also be functions of the scale factor perturbations $\delta S(s), s \le t$. Strictly speaking, while solving the Maxwell equations, we should take into account the current density produced by the matter fluid, ie, we should be solving the MHD equations in general relativity with the unperturbed electromagnetic fields being the ones that we have applied and the perturbations to these being obtained by approximately solving the MHD equations. More accurately, if the expansion of the universe is to be taken into consideration whilst doing this computation, we should assume that the background metric is the Robertson-Walker

$$d\tau^2 = dt^2 - S^2(t)dr^2/(1-kr^2) - S^2(t)r^2(d\theta^2 + sin^2(\theta)d\phi^2)$$

and then consider small metric perturbations $\delta g_{\mu\nu}(x)$ around this background metric. We should also consider small perturbations to the zero comoving velocity field that satisfies the geodesic equations for the Robertson-Walker spacetime. We should then set up the linearized Einstein-Maxwell equations for these metric perturbations with the electromagnetic field being small perturbations of the homogeneous and isotropic field corresponding to the cosmic microwave background radiation field plus the initial electromagnetic field applied locally in our laboratory and the velocity and density fields being small perturbations of the zero comoving field and the spatially constant density solution that satisfies the RW-metric based Einstein field equations. The resulting equations on solving would yield the spatio-temporally metric, velocity density and electromagnetic field perturbations. In particular we would obtain the electromagnetic field as a function of the scale factor $S(t)$ and the initial electromagnetic field from which we could in principle calculate the average electric and magnetic dipole moments using the method of classical statistical mechanics outlined above and hence derive the permittivity and permeability of the medium.

9. Applications of heat and mass transfer equations to cosmology
In the expanding universe, we write

$$v(t,r) = H(t)\mathbf{r} + \delta\mathbf{v}(t,\mathbf{r})$$

where $H(t) = S'(t)/S(t)$ is Hubble's constant and $\delta\mathbf{v}(t,\mathbf{r})$ is the inhomogeneous perturbation to the velocities that represent the velocity field of clumps of matter like galaxies. Likewise, we write

$$T(t,\mathbf{r}) = T_0(t) + \delta T(t,\mathbf{r})$$

and

$$\rho(t,\mathbf{r}) = \rho_0(t) + \delta\rho(t,\mathbf{r})$$

and substitute into the Navier-Stokes equations for heat and mass transfer:

$$\rho(\partial_t\mathbf{v}(t,\mathbf{r}) + (\mathbf{v},\nabla)\mathbf{v}(t,\mathbf{r}))$$

$$= -\nabla p(t,\mathbf{r}) - \rho(t,\mathbf{r})\nabla\delta\Phi(t,\mathbf{r}) + div(\sigma(t,\mathbf{r})) - - - (1)$$

$$\rho.[C_p\partial_t T(t,\mathbf{r}) + \mathbf{v},\nabla)T(t,\mathbf{r})]$$

$$= (\sigma,\nabla)\mathbf{v}(t,\mathbf{r}) + \nabla.(D\nabla T(t,\mathbf{r}))$$

where

$$\sigma_{ij} = \eta(v_{i,j} + v_{j,i})$$

and the equations of mass conservation and gravity:

$$div(\rho\mathbf{v}) + \partial_t \rho = 0,$$

$$\nabla^2 \delta\Phi(t, \mathbf{r}) = 4\pi G \delta\rho(t, \mathbf{r})$$

After linearizing these equations, we obtain differential equations which are first order in time for the variables $\delta\rho(t, \mathbf{r}), \delta\mathbf{v}(t, \mathbf{r})$ and $\delta T(t, \mathbf{r})$. If we wish in addition to take electromagnetic fields into consideration, then in the above Navier-Stokes equation (1), we must add an MHD term on the right

$$\mathbf{J} \times \mathbf{B}$$

where

$$\mathbf{J} = \sigma_0(\mathbf{E} + \mathbf{v} \times \mathbf{B})$$

with σ_0 representing the electrical conductivity of the matter fluid. These equations are to be solved along with the Maxwell equations. For the present, if we are not bothered about electromagnetic fields and we wish to quantize the above perturbation equations, then we must look at quantizing a differential equation of the form

$$d\xi(t)/dt = F(t, \xi(t)) - - - (2)$$

where $\xi(t)$ is a vector that at time t represents the velocity perturbations, the density perturbations and the temperature perturbations at the different spatial pixels that fill up the volume of the universe. An equation of the form (2) is generally not Hamiltonian as we could easily guess by looking at the temperature equation which contains a diffusion term. Thus, we cannot generally obtain (2) from Hamiltonian mechanics. We must instead supplement a Hamiltonian with Lindblad terms which we postulate arise owing to a connection between our universe and a surrounding bath.

10. Appendix: A derivation of the Lindlbad master equation

Let $H(t)$ denote the system Hamiltonian and $\rho(t)$ the density operator. It satisfies

$$i\rho'(t) = [H(t), \rho(t)]$$

which gives

$$i\rho(t+\tau) - i\rho(t) = \int_t^{t+\tau} [H(s_1), \rho(s_1)]ds_1$$

$$= \int_t^{t+\tau} [H(s_1), \rho(t) - i\int_t^{s_1} [H(s_2), \rho(s_2)]ds_2]ds_1$$

$$= [\int_t^{t+\tau} H(s_1)ds_1, \rho(t)] - i\int_{t<s_2<s_1<t+\tau} [H(s_1), [H(s_2), \rho(s_2)]ds_2 ds_1$$

$$\approx \tau[H(t), \rho(t)] - i \int_{t<s_2<s_1<t+\tau} [H(s_1), [H(s_2), \rho(t)]]ds_2 ds_1$$

From this equation, by assuming $H(s)$ to be a random Hermitian operator valued variable, it is easy to take averages and arrive at the GKSL equation. For example, if $(H_(s_1), H(s_2))$ takes the value (L_k, L_j) with probability $p(k,j)(s_1, s_2)$ so that $H(t)$ takes the value L_k with probability $p_k(s) = \sum_j p_{k,j}(s, s,')$, then assuming $\rho(t)$ to be independent of $H(s), s \geq t$, we get

$$< [H(t), \rho(t)] >= [\sum_k p_k(t)L_k, < \rho(t) >],$$

$$< [H(s_1), [H(s_2), \rho(s_2)]] >= \sum_{k,j} [p_{k,j}(s_1, s_2)(L_k L_j < \rho(s_2) >$$

$$+ < \rho(s_2) > L_k L_j - L_k < \rho(s_2) > L_j - L_j < \rho(s_2) > L_k]$$

which when substituted into the above equation and allowing $\tau \to 0$ so that $\tau.\sqrt{L_k} \to B_k$ results in the GKSL equation. It should be noted that although our derivation is based on a classical probabilistic averaging, we could easily extend this derivation to a quantum probabilistic averaging method by use of tensor products. Thus we approximate $\rho(t)$ by $\rho_S(t) \otimes \rho_B(t)$ where $\rho_S(t)$ is the system state and $\rho_B(t)$ the bath state and likewise,

$$H(t) = H_S(t) \otimes I_B + I_B \otimes H_B(t) + H_{SB}(t)$$

where $H_s(t)$ is the system Hamiltonian acting in the system Hilbert space, $H_B(t)$ is the bath Hamiltonian acting in the bath Hilbert space and $H_{SB}(t)$ is the interaction Hamiltonian acting in the tensor product of the system and bath Hilbert spaces. We leave it as an exercise to substitute these expressions into the above equation, take partial trace over the bath space and arrive at the GKSL equation for the system Hamiltonian.

11. Brief Highlights of this Research

[1] Derivation of the quantum Boltzmann equation by partial tracing the quantum Schrodinger-Liouville-Von-Neumann equation for N particles interacting with themselves and with external electromagnetic fields over other particles.

[2] Calculating the quantum averaged electric and magnetic dipole moments of a single particle using the one particle density operator obtained as an approximate solution to the quantum Boltzmann equation using perturbation theory.

[3] Specializing the computation of average dipole moments to Dirac Hamiltonians.

[4] From the quantum averaged electric and magnetic dipole moment per unit volume, derivation of the electric permittivity and magnetic permeability and hence the refractive index.

[5] Highlighting the dependence on temperature and the frequency of the electromagnetic field of the refractive index. Noting that temperature dependence arises due to solving the quantum Boltzmann equation with the initial density operator being the Gibbs state.

[6] Attempting to derive an equilibrium solution to the quantum Boltzmann equation in the presence of a static electromagnetic field by an iterative algorithm with the initialization of the algorithm being given by the Gibbs state.

[7] Applying statistical quantum field theory of infinite particle Fermionic systems to derive the temperature Green's function and from there the average dipole moments and thence the refractive index as a function of temperature and the electromagnetic field. The method is the standard Hartree-Fock method and works when we model our quantum liquid as an infinite particle Fermi system based on canonical second quantization of Fermi fields. This is superior to the quantum Boltzmann method since the latter is based on finite particle approximations using first quantization.

[8] An independent classical probabilisitic and quantum probabilistic derivation of the Lindblad master equation and its application to the the derivation of the quantum Boltzmann equation to obtain bath corrections to the refractive index.

[9] Derivation of the cosmological equations for the expanding universe and for the propagation of inhomogeneities in the form of matter field density, velocity and temperature fluctuations and electromagnetic field fluctuations from Newtonian particle mechanics and Eulerian fluid mechanics.

[10] Evaluation of the effects of cosmological expansion and the propagation of inhomogeneities on the measurement of refractive index.

[11] Hints about how to quantize the cosmological equations using the Lindblad formalism of open quantum systems.

[12] Hints about how to calculate the mean square fluctuations in the quantum electromagnetic field in an expanding universe with applications to its effect on the refractive index of materials.

12.Glossary of symbols

$A_\mu(x)$ Covariant components of the electromagnetic four potential.

$A^\mu(x)$ Contravariant components of the electromagnetic four potential.

$F_{\mu\nu}(x)$ Covariant components of the antisymmetric electromagnetic field tensor. $F_{0r} = -F_{r0}, r = 1,2,3$ are the electric field components while F_{12}, F_{23}, F_{31} are the magnetic field components.

$\rho(t)$ density matrix representing a mixed state of a quantum system at time t.

$\rho(t,r,r') =< r|\rho(t)|r' >$ position space representation of the density matrix of a mixed state of a quantum system.

$\rho_{12..N}(t)$ Joint mixed state of N particles of a quantum system.

$Tr_{23...N}\rho_{12...N}(t) = \rho_1(t)$ marginal mixed state of the first particle of an N particle quantum system. $Tr_{23...N}$ denotes the partial trace operation. In the position space representation,

$$[Tr_{23...N}\rho_{12..N}](t,r_1,r_1') = \int \rho_{12...N}(t,r_1,r_2...,r_N,r_1',r_2...,r_N)d^3r_2...d^3r_N$$

or equivalently in terms of countable orthonormal bases,

$$< e_{i_1}|[Tr_{23...N}\rho_{12...N}](t)|e_{j_1} >=$$

$$\sum_{i_2,...,i_N} < e_{i_1} \otimes e_{i_2} \otimes ... \otimes e_{i_N} |\rho_{12...N}| e_{j_1} \otimes e_{2,i_2} \otimes ... \otimes e_{N,i_N} >$$

Acknowledgements: I am grateful to Dr.Stephen A.Langford for several constructive suggestions based on his experimental data of refractive indices that helped me improve the content and the presentation of this paper. The data shown by Dr.Langford that helped in improving the content pertain to temperature and wavelength dependence of the refractive index of liquids.

13.References

[1] Landau and Lifshitz, "The classical theory of fields", Butterworth and Heinemann.

[2] Steven Weinberg, "Gravitation and Cosmology:Principles and Applications of the General Theory of Relativity", Wiley.

[3] Steven Weinberg, "The quantum theory of fields, vol.I", Cambridge University Press.

[4] Harish Parthasarathy, "Developments in mathematical and conceptual physics:Concepts and applications for engineers", Springer Nature 2020.

[5] K.R.Parthasarathy, "An introduction to quantum stochastic calculus", Birkhauser, 1992.

[6] M.Green, J.Schwarz, E.Witten, "Superstring Theory", Cambridge University Press.

[7] Stephen Arthur Langford, "Cumulative distribution functions for the refractive index of liquids as a function of temperature and wavelength", private communication.

[8] A.Fetter and Walecka, "Quantum Theory of Many Particle Systems", Dover, 1971.

[9] Richard Feynman, "The Feynman lectures on physics, vol.II"

[10] P.A.M.Dirac, "The principles of quantum mechanics", Oxford.

[11] Landau and Lifshitz, "Fluid mechanics", Butterworth and Heinemann.

For Product Safety Concerns and Information please contact our EU
representative GPSR@taylorandfrancis.com
Taylor & Francis Verlag GmbH, Kaufingerstraße 24, 80331 München, Germany

www.ingramcontent.com/pod-product-compliance
Lightning Source LLC
Chambersburg PA
CBHW060247220326
41598CB00027B/4015
* 9 7 8 1 0 3 2 4 0 5 1 4 8 *